LANDAU FERMI LIQUIDS
AND BEYOND

LANDAU FERMI LIQUIDS
AND BEYOND

V. Tripathi

CRC Press
Taylor & Francis Group
Boca Raton London New York

CRC Press is an imprint of the
Taylor & Francis Group, an **informa** business

CRC Press
Taylor & Francis Group
6000 Broken Sound Parkway NW, Suite 300
Boca Raton, FL 33487-2742

Printed on acid-free paper
Version Date: 20180119

International Standard Book Number-13: 978-1-4987-5543-6 (Hardback)

Visit the Taylor & Francis Web site at
http://www.taylorandfrancis.com

and the CRC Press Web site at
http://www.crcpress.com

To Poorvi and Charoo

Contents

List of Figures

Preface

The early phase of condensed matter physics was dominated by the development of our understanding, at the level of noninteracting particles, of the role quantum mechanics plays in the determining the properties of solids. Bloch's theorem, energy bands in crystals, Sommerfeld's independent electron theory and semiclassical treatments of electrons in metals and semiconductors all date from this period. The modern semiconductor industry owes its existence to this phase of condensed matter physics. The last sixty years have witnessed tremendous advancement in our understanding of the effect of interparticle interactions on the properties of solids. The early part of this phase saw the development of two important paradigms, the Landau Fermi liquid theory in 1956 and the Bardeen, Cooper, Schrieffer (BCS) theory of superconductivity in the following year. New approaches based on the many-body theory of interacting particles witnessed increasing use. The study of quantum mechanical effects of disorder led to new concepts such as the Kondo effect and the Anderson metal- insulator transition, and an understanding of the integer quantum Hall effect. The 1980s saw the discovery of high-Tc superconductivity and the fractional quantum Hall effect. The interparticle interactions in these strongly-correlated systems are much stronger than the particle kinetic energies, and conventional treatments based on many body techniques do not usually work here. Strongly-correlated systems is where many of the challenges in condensed matter physics currently lie. Understanding the effect of interacting particles in matter is thus still an ongoing revolution. The present book does not purport to discuss the interesting world of strongly-correlated electron systems. Instead, the emphasis is to consolidate in this book the concepts on weakly-interacting fermions where traditional perturbative many-body techniques are usually adequate. There already exist some classics on weakly-interacting Fermi systems. These books however do not cover one or more important later developments such as the Kondo effect, disorder effects and the Hertz theory of quantum criticality in metallic magnets. Part of the effort of this book is to bridge this gap and make it relevant to contemporary research trends.

This book grew out of lectures the author presented as part of two graduate-level courses at the Tata Institute of Fundamental Research. The audience largely consisted of graduate students engaged in research in experimental or theoretical condensed matter physics. This book targets graduate students and researchers who would like to learn about the phenomena related to weakly-interacting fermions in condensed matter physics as well as many-body concepts relevant to such systems.

The text is organized as follows. There are four parts: Phenomenology of Fermi Liquids, Microscopic Approach, Beyond Fermi Liquid Theory, and Open Questions. Part I, consisting of Chapters 1-5, provides a phenomenological treatment of the Landau Fermi liquid theory including Nozières' Fermi liquid theory for the low-temperature phase of the Kondo model. Part II (Microscopic Approach) contains three Chapters devoted to microscopic approaches to Fermi liquids. Chapters 6, 7 and 8 concern microscopic treatment of the response of ideal Fermi gases. Alongside standard topics such as Thomas-Fermi and Lindhard screening, Friedel and RKKY oscillations, Hartree and Hartree-Fock approximations, phonons in metals, and collective excitations, we also cover here some less discussed ones, namely Anderson orthogonality catastrophe, and the effect of disorder on electron interactions. Chapter 9 serves multiple purposes. First it provides a microscopic basis for Landau Fermi liquid phenomenology developed in Part I. Here we make explicit connections between the phenomenological Landau Fermi liquid parameters and the microscopic self-energy and vertex functions. Ward identities and their relation to response functions are discussed in some depth. A perturbative proof of Luttinger's theorem is provided. Second, we present an in-depth analysis of the Stoner instability, the Hertz theory for quantum critical metals, including the Doniach-Engelsberg mass enhancement phenomenon. Towards the end of this Chapter, we come across the first examples (in this book) of the breakdown of the Landau Fermi liquid theory. In Chapter 9, we assume the reader has some familiarity with perturbative techniques for many-particle systems. Part III (Beyond Fermi Liquid Theory) covers two important routes leading to the breakdown of the Landau Fermi liquid paradigm staying within the realm of perturbation theory: transverse electromagnetic fluctuations and overscreening in the Kondo model. Part IV contains a discussion of some open questions in the area.

The author would like to express his deep gratitude to Dima Khmelnitskii for his invaluable advice on the selection of topics for this book. A special thanks also to Deepak Dhar for his suggestions to improve the manuscript.

Mumbai

Part I

Phenomenology of Fermi Liquids

Chapter 1

Quasiparticle Concept

1.1 Noninteracting fermions

Fermions obey Pauli's exclusion principle that no two of them can occupy the same quantum state. If the fermions are noninteracting, the Pauli principle immediately allows us to factorize the grand partition function (hereafter referred to as simply the partition function),

$$\mathcal{Z} = \sum_{N,n} e^{-(\epsilon_n(N)-\mu N)/k_B T}, \tag{1.1}$$

into the following product over the single particle eigenstates of the system:

$$\mathcal{Z} = \prod_i \left(1 + e^{-(\epsilon_i^{(0)}-\mu)/k_B T}\right). \tag{1.2}$$

Here $\epsilon_n(N)$ is the energy eigenvalue for the nth state of the system with N fermions, $\epsilon_i^{(0)}$ is the energy of the ith *single particle* state, and μ is the chemical potential. Likewise, the thermodynamic potential,

$$\Omega = -k_B T \ln \mathcal{Z},$$

also decomposes into a sum over the thermodynamic potentials of the single particle states. The chemical potential is related to the expectation value of the number of particles, $\langle N \rangle = -\partial\Omega/\partial\mu = -\sum_i \partial\Omega_i^{(0)}/\partial\mu$, in the system. Thus for the ith single-particle state, we may introduce a distribution function,

$$n_i^{(0)} = -\frac{\partial \Omega_i^{(0)}}{\partial \mu} = \frac{1}{e^{(\epsilon_i^{(0)} - \mu)/k_B T} + 1}, \tag{1.3}$$

such that $\langle N \rangle = \sum_i n_i^{(0)}$. Eq.(1.3) is the well-known Fermi-Dirac distribution.

For any given value of $\langle N \rangle$, the chemical potential necessarily is a temperature-dependent function. Suppose that μ were independent of temperature. Then $n_i^{(0)} \approx 1/2$ as long as $|\epsilon_i^{(0)} - \mu| \ll k_B T$, and $\sum_i n_i^{(0)}$ will eventually exceed $\langle N \rangle$ at high enough temperatures. Clearly, for preserving $\langle N \rangle$, μ must decrease with increasing temperature, and ultimately, at high enough temperatures, $\mu/k_B T$ should acquire large and negative values. At such high temperatures, the distribution function, Eq.(1.3), can be approximated as

$$n_i^{(0)} \approx e^{-(\epsilon_i^{(0)} - \mu)/k_B T}. \tag{1.4}$$

Eq.(1.4) is the familiar Maxwell-Boltzmann (MB) distribution obeyed by *distinguishable* particles. Thus at high enough temperatures, the fact that fermions obey the Pauli exclusion principle is not important. Physically, the effect of the Pauli exclusion principle is felt only when the occupation probability of a given quantum state is of the order of unity. In the MB case, the probability of occupation of any quantum state is always small and hence Pauli's principle is not important.

For an ideal, free MB gas, $\epsilon_{\mathbf{p}}^{(0)} = \mathbf{p}^2/2m$. Using the relation

$$\langle N \rangle = \frac{gV}{(2\pi\hbar)^3} \int d^3\mathbf{p}\, e^{-(\mathbf{p}^2/2m - \mu)/k_B T},$$

we find that for the ideal, free MB gas, the chemical potential is

$$\mu = k_B T \ln(n\lambda_T^3/g), \tag{1.5}$$

where $n = \langle N \rangle / V$ is the number density of the particles, g is the degeneracy of the energy eigenstates ($g = 2$ for spin-1/2 fermions), and

$$\lambda_T = \sqrt{\frac{2\pi\hbar^2}{mk_B T}} \tag{1.6}$$

is the thermal wavelength of the particles. λ_T is of the order of the de Broglie wavelength of a free particle with kinetic energy $\mathbf{p}^2/2m \sim (3/2)k_B T$.

Note that the validity of approximating the Fermi-Dirac distribution by the Maxwell-Boltzmann distribution rested on satisfying the condition that $\mu/k_B T$ is a large and negative number. This requirement is satisfied [see Eq.(1.5)] as long as $n\lambda_T^3 \ll 1$. At large particle density n and at low enough temperatures, $n\lambda_T^3$ will no longer remain a small quantity and the Maxwell-Boltzmann description will no longer be valid. We must then use the Fermi-Dirac distribution. Below the temperature

$$T_d = \frac{2\pi\hbar^2}{mk_B} n^{2/3},$$

known as the degeneracy temperature, the classical description is inappropriate. Physically, at such low temperatures, the Maxwell-Boltzmann occupation probability of the lower quantum states may approach unity, and this is precisely the situation where we need to worry about the Pauli exclusion principle.

Next we consider a Fermi gas at temperatures much below the degeneracy temperature T_d. At $T = 0$, the Fermi-Dirac distribution is a step function,

$$n_i^{(0)}(T = 0) = \Theta(\epsilon_F - \epsilon_i^{(0)}), \tag{1.7}$$

and ϵ_F, referred to as the Fermi energy, is the chemical potential at $T = 0$. $\epsilon_i^{(0)} = \epsilon_{\mathbf{p}}^{(0)} = \mathbf{p}^2/2m$ for a free Fermi gas. Using the relation

$$n = \frac{g}{(2\pi\hbar)^3} \int d^3\mathbf{p}\, n_{\mathbf{p}}^{(0)}(T), \tag{1.8}$$

for the number density n of the particles together with Eq.(1.7), we obtain an expression for the Fermi energy of a gas of free fermions:

$$\epsilon_F = \frac{\hbar^2}{2m}(6\pi^2 n/g)^{2/3}. \tag{1.9}$$

The Fermi energy is of the order of $k_B T_d$, so in the Fermi liquid literature ϵ_F is usually regarded as the energy scale below which the Pauli principle is important. In the remaining sections, we will regard the Fermi temperature $T_F = \epsilon_F/k_B$ as the degeneracy temperature of the gas of fermions. ϵ_F is the energy of the highest occupied state at $T = 0$. In typical metals, T_F has fairly high values of the order of 10^4K.

At finite but low temperatures, one can show, using Eq.(1.8) and Eq.(1.9), that the chemical potential decreases as

$$\mu(T) = \epsilon_F \left[1 - \frac{\pi^2}{12}\left(\frac{k_B T}{\epsilon_F}\right)^2 + \cdots\right]. \tag{1.10}$$

1.1.1 Low energy properties of a Fermi gas

We have seen that the ground state configuration of a Fermi gas consists of completely filled energy levels up to ϵ_F, and unoccupied levels for energy greater than ϵ_F. At a finite temperature, some of the fermions in the occupied

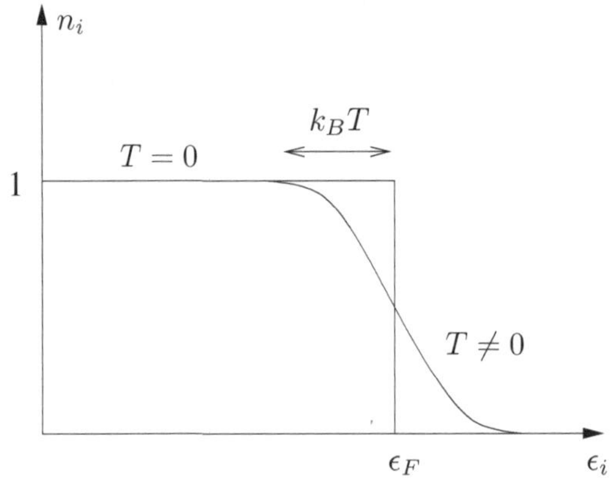

FIGURE 1.1: The degenerate Fermi gas at absolute zero and at $T \ll \epsilon_F/k_B$. The number density of excitations is proportional to the area of the two approximately triangular regions.

levels will be excited to the unoccupied levels above ϵ_F. The properties of a degenerate Fermi gas were first studied by Sommerfeld in 1927. They turn out to be very different from a classical ideal gas.

In an ideal MB gas, the average kinetic energy of a particle is $(3/2)k_BT$ and the specific heat is $C_V = (3/2)nk_B$. Consider now a Fermi gas far below the degeneracy temperature. From Eq.(1.3) we find that when $T \ll \epsilon_F/k_B \equiv \mu(T = 0)/k_B$, the distribution function deviates from the step function only in a narrow range of energies where the exponential is of order unity, that is, $|\epsilon_i^{(0)} - \epsilon_F| \leq k_BT$. The number of excitations per unit volume for a free degenerate Fermi gas (see Figure 1.1) is

$$\delta n_{\text{exc}} \approx \frac{1}{2} k_BTg(\epsilon_F),$$

where

$$g(\epsilon) = \frac{g}{4\pi^2}\left(\frac{2m}{\hbar^2}\right)^{3/2} \epsilon^{1/2} = \frac{gmp}{2\pi^2\hbar^3} \tag{1.11}$$

is the density of states of free particles at energy ϵ. Using Eq.(1.9) and Eq.(1.11), we obtain $g(\epsilon_F) = 3n/4\epsilon_F$, and

$$\frac{\delta n_{\text{exc}}}{n} \sim \frac{k_BT}{\epsilon_F} \ll 1.$$

Thus a very small fraction of the fermions gets excited. As the energy of these excitations is of the order of k_BT, the average excitation energy per particle in the degenerate Fermi gas is about $(k_BT/\epsilon_F)k_BT$, which is much smaller than the MB value. The specific heat has a linear$-T$ dependence, $C_V \sim nk_B(k_BT/\epsilon_F)$. A more accurate calculation gives

$$C_V = \gamma T = \frac{mp_F k_B^2}{3\hbar^3} T \equiv \frac{\pi^2}{2} nk_B \frac{k_BT}{\epsilon_F}. \tag{1.12}$$

In the presence of a small external magnetic field B_{ext} such that $\mu_{B,N}B_{\text{ext}} \ll k_BT$, the fermions show a paramagnetic response on account of their spin, and if they are charged, an additional diamagnetic response due to their spatial motion. Here $\mu_{B,N}$ is the Bohr (or nuclear) magneton depending on whether the spin-1/2 is of electronic or nuclear origin. A system of chargeless spin-1/2 fermions will acquire a magnetization

$$\begin{aligned} M &= -(1/V)(\partial\Omega/\partial B_{\text{ext}})_{T,V,\mu} = -\mu_{B,N}(\partial^2\Omega/\partial\mu^2)(\mu_{B,N}B_{\text{ext}}/V) \\ &= (\mu_{B,N}^2/V)(\partial N/\partial\mu)_{T,V}B_{\text{ext}} \equiv \chi B_{\text{ext}} \end{aligned} \tag{1.13}$$

along the direction of the field. Now for a Maxwell-Boltzmann gas, we can use the expression for the chemical potential in Eq.(1.5); this gives us $\chi = (n\mu_{B,N}^2/k_BT)$, which is the Curie susceptibility for free spins. In comparison, the magnetization of a degenerate spin-1/2 Fermi gas is

$$M = \delta n_\uparrow - \delta n_\downarrow = \mu_{B,N}(\mu_{B,N}B_{\text{ext}})g(\epsilon_F),$$

which leads to a temperature-independent susceptibility

$$\chi = \frac{3n}{2}\frac{\mu_B^2}{\epsilon_F} \equiv \frac{3}{\pi^2}\frac{\gamma}{k_B^2}\mu_B^2. \tag{1.14}$$

Here we used the relation $g(\epsilon_F) = 3n/2\epsilon_F = (1/V)(\partial N/\partial\mu)_{T,V}$ for $k_BT \ll \epsilon_F$. χ in Eq.(1.14) is referred to as the Pauli spin susceptibility. The ratio of the coefficient of linear specific heat, γ, to the Pauli spin susceptibility at $T = 0$ is a universal constant. At finite temperatures, the Pauli susceptibility may be obtained from Eq.(1.13) and the temperature dependence of the chemical potential, Eq.(1.10).

The diamagnetic susceptibility χ_{dia} is less straightforward to calculate and requires dealing with Landau levels. The result is that $\chi_{\text{dia}} = -(1/3)\chi$ for both the Maxwell-Boltzmann gas as well as the degenerate (and charged) Fermi gas[1]. The expression for the degenerate Fermi gas is known as Landau diamagnetic susceptibility.

Experimentally, until recently, the only known Fermi gas[2] was a dilute solution of ^{3}He dissolved in superfluid ^{4}He. Now it has become possible to produce nearly ideal degenerate Fermi gases in cold atom traps. One of the earliest reports was, by DeMarco and Jin in 1999, of the production of a degenerate Fermi gas composed of ^{40}K atoms at a temperature of about 1μK.

[1] L. Landau, *Zeitschrift für Physik* **64**, 629 (1930).

[2] Electrons in metals do not strictly fall in this category on account of their long-range Coulomb interaction.

1.2 Quasiparticles

In Sec. 1.1 we studied some thermodynamic properties of the noninteracting Fermi gas below the degeneracy temperature. Our starting point was the partition function of Eq.(1.1) and Eq.(1.2). We found that the properties are essentially determined by a small fraction ($\sim k_BT/\epsilon_F$) of the fermions within an energy range of the order of k_BT at the Fermi energy. A natural question arises as to how the properties of a degenerate Fermi gas would change if the fermions interacted mutually. Unlike the noninteracting case where we can say that the total energy is just the sum of the energies of the individual particles, it is impossible in general to determine the energy levels of a large number of interacting particles. If the interaction is weak, then the interaction may be regarded as a perturbation and the corrections to thermodynamic properties may be estimated. Frequently, however, the interaction is not small and we cannot employ the same approach as the noninteracting system.

In a series of papers beginning in 1956, Landau developed a phenomenological theory of degenerate interacting fermions, or, a *Fermi liquid.* [3] Landau's Fermi liquid theory has been a tremendous success and has had a great influence on our understanding of a number of condensed matter systems, whether it is the properties of liquid ^{3}He, electrons in metals and typical semiconductors, the physics of magnetic impurities in metals (Kondo effect), or more exotic systems such as heavy fermion metals and fermions in cold atom traps.

^{3}He and neutral fermions in cold atom traps are perhaps the only interacting Fermi systems we encounter commonly, for which the Landau Fermi liquid theory works very well. The applicability of the theory is somewhat less for electrons in metals and semiconductors because of extra complications due to the lattice and electromagnetic interactions.

Despite the fact that we are dealing with a many-particle system with strong quantum effects due to Pauli's exclusion principle, and often significant interparticle interactions, the properties of Fermi liquids are better understood than the corresponding Maxwell-Boltzmann case. Central to this success is the concept of elementary excitations that Landau introduced. As $T \to 0$, the most important states in the partition function are the weakly excited states, or elementary excitations, whose energies are close to the ground state. The nature of these excited states can be fairly well understood by using very general considerations (such as symmetries) that are valid regardless of the magnitude and specific details of the interaction. From a knowledge of the low-lying energy levels, we can calculate various thermodynamic properties.

A basic assumption made by Landau was that to a first approximation, the structure of the energy levels of the interacting system follows the same

[3] We mean, strictly speaking, a normal Fermi liquid where the properties show a qualitative similarity with the Fermi gas irrespective of the strength of interactions, such as having well-defined Fermi surface, a linear$-T$ specific heat, etc.

principles as that of the non-interacting counterpart. In other words, the assumption is that any energy level of the interacting system can be obtained as the sum of the energies of a certain number of "quasiparticles". In free space, the quasiparticles may be labeled with momentum $\mathbf{p}$ and energy $\epsilon_{\mathbf{p}}$, and $\epsilon_{\mathbf{p}}$ in general will be different from its noninteracting counterpart $\epsilon_{\mathbf{p}}^{(0)} = p^2/2m$.

Let us turn to an actual example. Consider first a noninteracting Fermi gas at $T = 0$. As we have seen in Sec. 1.1, the ground state configuration consists of completely filled states up to an energy ϵ_F, or, in momentum space, a filled sphere with radius equal to the Fermi momentum

$$p_F = \hbar(3\pi^2 n)^{1/3}. \tag{1.15}$$

In the excited state, the momentum distribution of the particles will be different. Clearly, every excited state can be constructed by moving a number of particles from the interior of the Fermi sphere to the exterior, and each such elementary act leads to the creation of a particle with $p > p_F$ and a "hole" with $p < p_F$.[4] The excitation energy of the system is simply additive,

$$\delta E = g \sum_{\mathbf{p}} \epsilon_{\mathbf{p}}^{(0)} \delta n_{\mathbf{p}}. \tag{1.16}$$

Here $\delta n_{\mathbf{p}} = n_{\mathbf{p}} - n_{\mathbf{p}}^{(0)}$ is the change in occupation number of the state with momentum $\mathbf{p}$. These particles and holes can be regarded as the elementary excitations of the Fermi gas. The weakly excited states have nonzero $\delta n_{\mathbf{p}}$ only in the vicinity of p_F; hence, we normally measure the energy of the elementary excitations from the surface of the Fermi sphere. Particle-type excitations have an energy

$$\xi_{\mathbf{p}} = \frac{p^2}{2m} - \frac{p_F^2}{2m} \approx v_F(p - p_F), \tag{1.17}$$

and hole-type excitations have an energy

$$-\xi_{\mathbf{p}} \approx v_F(p_F - p). \tag{1.18}$$

Consider now a Fermi liquid. Landau assumed that the ground state and weakly excited states of a Fermi liquid closely resemble the ground state and weakly excited state of a Fermi gas, and can therefore be described by elementary excitations with spin-1/2 and momenta near p_F. The reasoning goes as follows. Suppose we begin with a Fermi gas and very slowly turn on interparticle interaction. Under such adiabatic perturbation, the original eigenstates will slowly evolve into eigenstates of the interacting system. Of course, it does not have to be this way. For example, a phase transition, such as superconductivity, may occur that will lead to a complete reorganization of the ground

[4] Excitations in nature are either of Bose type (integer spin) or Fermi type (half-integer spin). A basic law of quantum mechanics is that transitions can only occur between states differing by an integer spin. Therefore, Fermi type excitations always appear in pairs.

state. Landau's assumption can therefore be considered as part of the definition of a normal Fermi liquid. Also, if we add a particle with momentum $\mathbf{p}$ to a Fermi gas, then turning on any momentum-conserving interaction will not change the total momentum of the system. This elementary excitation can therefore can be labeled with the same momentum $\mathbf{p}$ in the interacting case. Thus there is a one-to-one correspondence between the momentum states of the free Fermi gas and the free Fermi liquid. This implies that in the Fermi liquid, the momentum of the highest occupied state is also equal to p_F, the noninteracting value.

We believe this is the appropriate moment to introduce Landau's concept of the quasiparticle for the Fermi liquid. Imagine filling one by one the momentum states of a Fermi liquid with some hypothetical quasiparticles up to a maximum value p_F. Because of the above one-to-one correspondence, the number of such quasiparticles will be equal to the number of fermions. Exciting the Fermi liquid produces an equal number of particles and holes. So, the total number of quasiparticles is conserved.

The above similarities notwithstanding, there are important differences between a Fermi gas and a Fermi liquid because the excitations in the liquid interact with each other. Superconductivity is the most striking phenomenon that would not have otherwise occurred in the absence of interaction of the electrons in a metal. Crucial differences are to be found even in the normal state. For a Fermi gas, the quasiparticle description is valid for excitations of arbitrary energy compared to ϵ_F. The quasiparticle description makes sense only if the relaxation rate (i.e., inverse of the lifetime τ) of the quasiparticles is much smaller than their energy. This is evidently not a problem in a Fermi gas where the quasiparticle states are the eigenstates of the system, and, therefore, infinitely long-lived. We now show that in a Fermi liquid, the quasiparticle description is valid only for weakly excited states because the relaxation rate will exceed the quasiparticle energy except for excitations with momenta near p_F.

1.2.1 Quasiparticle lifetime

The relaxation of quasiparticle excitations is due to their mutual collisions and consequent decay into several others. Since the numbers of particles and holes in the system are equal, a particle will decay into at least two particles and a hole. Crudely speaking, the relaxation rate, $\tau_{\mathbf{p}1}^{-1}$, of a particle with momentum p_1 close to the Fermi momentum p_F is proportional to the phase space available for the decay process. The available phase space is determined by energy and momentum conservation:

$$\tau_{\mathbf{p}1}^{-1} \propto \int d\mathbf{p}_2 d\mathbf{p}_3 \, \delta(\epsilon_{\mathbf{p}1} + \epsilon_{\mathbf{p}2} - \epsilon_{\mathbf{p}3} - \epsilon_{\mathbf{p}4}), \tag{1.19}$$

where $\mathbf{p}_2$, ($p_2 < p_F$) is the momentum of the hole and $\mathbf{p}_3$, ($p_3 > p_F$), and $\mathbf{p}_4$, ($p_4 = |\mathbf{p}_1 + \mathbf{p}_2 - \mathbf{p}_3| > p_F$) are the momenta of the daughter particles. Since $\mathbf{p}_3$ and $\mathbf{p}_4$ represent particles, their momenta are at least p_F each. Also because of conservation of energy, $\epsilon_{\mathbf{p}3}$ cannot exceed $\epsilon_{\mathbf{p}1} + \epsilon_{\mathbf{p}2} - \mu(0)$. Here $\mu(0)$ is the chemical potential at $T = 0$. For noninteracting fermions, $\mu(0) = \epsilon_F$. This gives us a constraint on $\mathbf{p}_3$ in relation to $\mathbf{p}_2$:

$$\mu(0) < \epsilon_{\mathbf{p}3} < \epsilon_{\mathbf{p}1} + \epsilon_{\mathbf{p}2} - \mu(0). \tag{1.20}$$

Near the Fermi surface, the dispersions $\epsilon_{\mathbf{p}}$ can be linearized as in Eq.(1.17) and Eq.(1.18), so we can replace $\epsilon_{\mathbf{p}} - \mu(0)$ by $v_F(p - p_F)$ everywhere. That simplifies the constraint in Eq.(1.20) to

$$p_F < p_3 < p_1 + p_2 - p_F.$$

We need a constraint on $\mathbf{p}_2$ in relation to $\mathbf{p}_1$. Again using the fact that $\mathbf{p}_3$ and $\mathbf{p}_4$ represent particles, the total energy of quasiparticles with labels $\mathbf{p}_1$ and $\mathbf{p}_2$ is at least $2\mu(0)$; thus,

$$2\mu(0) - \epsilon_{\mathbf{p}1} < \epsilon_{\mathbf{p}2} < \mu(0),$$

or upon linearizing,

$$2p_F - p_1 < p_2 < p_F.$$

Thus $\mathbf{p}_2$ lies just inside the Fermi sphere within a narrow shell of thickness $p_1 - p_F \ll p_F$. Similarly the first constraint shows (in combination with the constraint on $\mathbf{p}_2$) that $\mathbf{p}_3$ lies just outside the Fermi sphere in a narrow shell. So far we have constraints on the magnitudes but not the directions of $\mathbf{p}_2$ and $\mathbf{p}_3$. $\mathbf{p}_2$ can have any orientation with respect to $\mathbf{p}_1$. The orientation of $\mathbf{p}_3$ with respect to $\mathbf{p}_1 + \mathbf{p}_2$ is limited by the energy conservation. Figure 1.2 shows one instance where the orientation $\mathbf{p}_3$ is such that the energy conservation constraint cannot be satisfied.

Integration over the angle between $\mathbf{p}_3$ and $\mathbf{p}_1 + \mathbf{p}_2$ in Eq.(1.19) gets rid of the delta function enforcing energy conservation. The remaining integrations in the narrow shell ($p_2 \approx p_3 \approx p_F$) about the Fermi surface give the factor $(p_1 - p_F)^2$, thus,

$$\tau_{\mathbf{p}1}^{-1} \propto (p_1 - p_F)^2. \tag{1.21}$$

On the other hand, the energy of the excitation is proportional to $(p_1 - p_F)$. Therefore, the relaxation rate of the quasiparticle excitation is small compared to the excitation energy only in the vicinity of the Fermi surface. At a small but finite temperature, the energy of an excitation is of the order of T. Since the quasiparticle relaxation rate $\tau_{\mathbf{p}}^{-1}$ is proportional to the square of the excitation energy, it follows that $\tau_{\mathbf{p}}^{-1}$ increases as T^2 with increasing temperature. Higher order processes, where a particle near the Fermi surface decays into $2m + 1$ quasiparticles ($m > 1$), will lead to extra attenuation $\propto (p_1 - p_F)^{2m}$, and these can be neglected in comparison to Eq.(1.21).

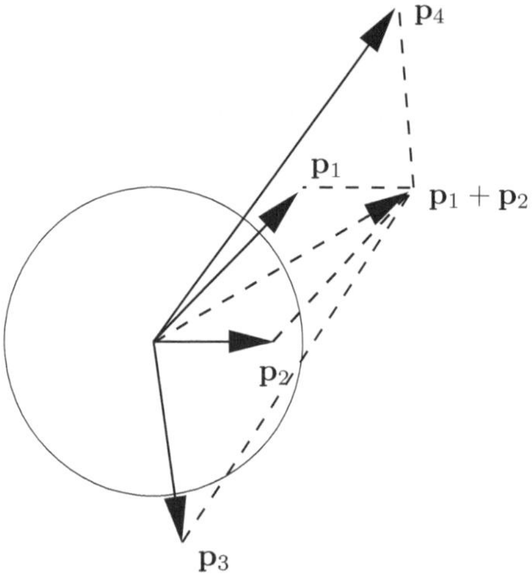

FIGURE 1.2: The orientation of $\mathbf{p}_3$ with respect to $\mathbf{p}_1 + \mathbf{p}_2$ shown here conserves momentum but not energy as $\epsilon_{\mathbf{p}3} + \epsilon_{\mathbf{p}4} > \epsilon_{\mathbf{p}1} + \epsilon_{\mathbf{p}2}$. The circle has a radius p_F and denotes the Fermi sphere.

1.2.2 Quasiparticle energy and interaction

Apart from the assumptions made so far regarding the nature of the quasiparticles, the Landau theory makes one more assumption regarding the interaction of quasiparticles – that the interaction can be described by a self-consistent field acting on one quasiparticle due to the other quasiparticles. The energy of the system cannot then be expressed as the sum of the energies of the separate quasiparticles. Instead, the energy is a functional of their distribution functions. In a manner akin to Eq.(1.16) for noninteracting fermions, we write (taking $g = 2$ and the continuum limit),

$$\delta E[n_{\mathbf{p}}] = 2V \int \frac{d\mathbf{p}}{(2\pi\hbar)^3}\, \epsilon_{\mathbf{p}} \delta n_{\mathbf{p}}. \tag{1.22}$$

From this definition, we gather that $\epsilon_{\mathbf{p}}$ is the change in energy of the system upon adding one quasiparticle with momentum $\mathbf{p}$. Also the deviation $\delta n_{\mathbf{p}}$ is no longer measured from the distribution $n_{\mathbf{p}}^{(0)}$ of the ideal Fermi gas. Rather, it is measured from the (yet unknown) distribution function of the Fermi liquid. In presence of magnetic interactions or a magnetic field, the distribution function as well as the quasiparticle energy should be regarded as having spin indices, thus,

$$\delta(E/V) = \sum_{\alpha,\beta} \int \frac{d\mathbf{p}}{(2\pi\hbar)^3}\, \epsilon_{\mathbf{p},\alpha\beta}\, \delta n_{\mathbf{p},\beta\alpha} \equiv \mathrm{Tr}_\sigma \int (d\mathbf{p})\, \epsilon_{\mathbf{p}}(\boldsymbol{\sigma})\, \delta n_{\mathbf{p}}(\boldsymbol{\sigma}). \tag{1.23}$$

Similarly we have for the entropy of of the quasiparticles,

$$S/V = -\mathrm{Tr}_\sigma \int (d\mathbf{p})\, [n \ln n + (1-n)\ln(1-n)]. \tag{1.24}$$

Maximizing the entropy S subject to the constraints $\delta N = \delta E = 0$ leads to the distribution function for the quasiparticles:

$$n_{\mathbf{p}} = \frac{1}{e^{(\epsilon_{\mathbf{p}}-\mu)/k_B T} + 1}. \tag{1.25}$$

The difference from the noninteracting case is that $\epsilon_{\mathbf{p}}$ itself is a functional of $n_{\mathbf{p}}$ and depends on the temperature as well. So Eq.(1.25) is actually more complicated than it appears at first sight. Let $\epsilon_{\mathbf{p}}^{(0)}$ henceforth denote the equilibrium energy of the quasiparticles at $T = 0$.[5]

In general, for a Fermi liquid, $\epsilon_{\mathbf{p}}^{(0)}$ will not be given by $p^2/2m$. In Eq.(1.25), we can expand the quasiparticle energy in a Taylor series in δn in the vicinity of $\epsilon_{\mathbf{p}}^{(0)}$:

$$\begin{aligned} \epsilon_{\mathbf{p},\alpha\beta} &= \epsilon_{\mathbf{p},\alpha\beta}^{(0)} + \delta\epsilon_{\mathbf{p},\alpha\beta} \\ &= \epsilon_{\mathbf{p},\alpha\beta}^{(0)} + \sum_{\gamma,\delta} \int (d\mathbf{p}')\, f_{\alpha\beta,\gamma\delta}(\mathbf{p},\mathbf{p}')\delta n_{\mathbf{p}',\gamma\delta}. \end{aligned} \tag{1.26}$$

More compactly,

$$\epsilon_{\mathbf{p}}(\boldsymbol{\sigma}) = \epsilon_{\mathbf{p}}^{(0)}(\boldsymbol{\sigma}) + \mathrm{Tr}_{\boldsymbol{\sigma}'} \int (d\mathbf{p}')\, f(\mathbf{p},\boldsymbol{\sigma};\mathbf{p}',\boldsymbol{\sigma}')\delta n_{\mathbf{p}'}(\boldsymbol{\sigma}'). \tag{1.27}$$

Typically, $f(\mathbf{p},\mathbf{p}')$ will be of the order of $1/(N/V)$ as can be seen from the following argument. The interaction term in Eq.(1.26) is of the order of $C(N/V) \times (N_{\mathrm{qp}}/N)$, where $N_{\mathrm{qp}} = V \int (d\mathbf{p}')|\delta n_{\mathbf{p}'}|$ is the number of quasiparticle *excitations* and C is some constant characterizing the strength of interaction. Now (N_{qp}/N) is small for that is a requirement for the validity of the Landau theory, but, it is still of order one, i.e., $(N/V)^0$. We thus require that C, and therefore, f, decrease faster than $1/(N/V)$; otherwise the interaction term will diverge at large fermion densities. Eq.(1.26) is perhaps the most crucial element of the Landau Fermi liquid theory. Its important feature is the second term that would have been absent if the quasiparticles were noninteracting. We will see later that the function f is related to the forward scattering amplitude of two quasiparticles. Here we make a note that the dimension of f is energy $\times$ volume, or, inverse of density of states. The quantity $f(\mathbf{p},\boldsymbol{\sigma};\mathbf{p}',\boldsymbol{\sigma}')\delta n_{\mathbf{p}'}(\boldsymbol{\sigma}')$ is known in the literature as the Landau f-function.

The spin-dependence of f can be generally written as

$$f(\mathbf{p},\boldsymbol{\sigma};\mathbf{p}',\boldsymbol{\sigma}') = \varphi(\mathbf{p},\mathbf{p}') + \sum_{i,k} \sigma_i\sigma_k\zeta_{ik}(\mathbf{p},\mathbf{p}'), \tag{1.28}$$

[5] We would not like to have too many notations for different excitation energies. Some symbols will be re-used if that causes no conflict.

where σ_i are the Pauli spin matrices, which are related to the fermion spin operator, $\mathbf{s}$, through $\mathbf{s} = (1/2)\boldsymbol{\sigma}$.

If f is invariant under time reversal symmetry, as would be the case in the absence of a magnetic field, the following condition is satisfied:

$$f(\mathbf{p}, \boldsymbol{\sigma}; \mathbf{p}', \boldsymbol{\sigma}') = f(-\mathbf{p}, -\boldsymbol{\sigma}; -\mathbf{p}', -\boldsymbol{\sigma}'). \tag{1.29}$$

If, in addition, the Fermi surface is invariant under $(\mathbf{p} \to -\mathbf{p})$, we have

$$f(\mathbf{p}, \boldsymbol{\sigma}; \mathbf{p}', \boldsymbol{\sigma}') = f(\mathbf{p}, -\boldsymbol{\sigma}; \mathbf{p}', -\boldsymbol{\sigma}'). \tag{1.30}$$

In other words, $f(\mathbf{p}, \boldsymbol{\sigma}; \mathbf{p}', \boldsymbol{\sigma}')$ depends only on the relative orientation of the spins. Exchange interaction, being electrostatic in origin, is one example that satisfies time reversal symmetry. In this case the spin part will have a simpler form, $(\boldsymbol{\sigma} \cdot \boldsymbol{\sigma}')\zeta(\mathbf{p}, \mathbf{p}')$.

There are only two independent components of f if Eq.(1.30) is satisfied. Instead of using φ and ζ, we can also write in terms of spin symmetric and antisymmetric components:

$$\begin{aligned} f^{\uparrow\uparrow}(\mathbf{p}, \mathbf{p}') &= f^s(\mathbf{p}, \mathbf{p}') + f^a(\mathbf{p}, \mathbf{p}'), \\ f^{\uparrow\downarrow}(\mathbf{p}, \mathbf{p}') &= f^s(\mathbf{p}, \mathbf{p}') - f^a(\mathbf{p}, \mathbf{p}'). \end{aligned} \tag{1.31}$$

Comparing with our earlier notation for the spin part, $(\boldsymbol{\sigma} \cdot \boldsymbol{\sigma}')\zeta(\mathbf{p}, \mathbf{p}')$, one can check that

$$\zeta(\mathbf{p}, \mathbf{p}') = f^a(\mathbf{p}, \mathbf{p}'). \tag{1.32}$$

The φ, ζ notation is usually encountered in the Russian literature. We will be using the notation $f = f^s + (\boldsymbol{\sigma} \cdot \boldsymbol{\sigma}')f^a$.

As the relevant quasiparticle momenta lie in a narrow shell at the Fermi wavevector, the momentum dependence of $f(\mathbf{p}, \mathbf{p}')$ is primarily through the orientations of $\mathbf{p}$ and $\mathbf{p}'$. In an isotropic system, $f(\mathbf{p}, \mathbf{p}')$ will depend not on the individual orientations of the momenta, but on the angle χ between their orientations. In that case we can expand f in a series of Legendre polynomials:

$$f^{s(a)}(\mathbf{p}) = \sum_{l=0}^{\infty} f_l^{s(a)} P_l(\cos\chi). \tag{1.33}$$

1.3 Thermodynamic properties

1.3.1 Effective mass and specific heat

In the absence of a magnetic field, the energy of the quasiparticles does not depend upon the spin. Then $\epsilon^{(0)}$ in Eq.(1.26) will depend only on p. We can expand $\epsilon^{(0)}$ in a series around the Fermi momentum:

$$\xi_p = \epsilon_{\mathbf{p}}^{(0)} - \mu(0) = v(p - p_F). \tag{1.34}$$

Note that p_F in Eq.(1.34) is the same as the noninteracting value, Eq.(1.15). This we have justified in Sec. 1.2. v is the velocity of the excitations at the Fermi surface. In analogy with the noninteracting case, we *define* an effective mass m^* :

$$v = \frac{p_F}{m^*}. \tag{1.35}$$

We now derive a relation between the effective mass m^* and the Fermi liquid parameter f. First note that the momentum $\mathbf{P}$ of a given volume V of the Fermi liquid is equal to the momentum of the quasiparticles in the volume, that is,

$$\mathbf{P} = 2V \int (d\mathbf{p})\, \mathbf{p}\, n_{\mathbf{p}}. \tag{1.36}$$

We have also illustrated before Landau's argument that the number density of quasiparticles and fermions is the same. Therefore, the *fermion* current density $\mathbf{j}$ is equal to the quasiparticle current density, i.e.,

$$\mathbf{j} = 2 \int (d\mathbf{p})\, \mathbf{v}_{\mathbf{p}}\, n_{\mathbf{p}}, \tag{1.37}$$

where $\mathbf{v}_{\mathbf{p}}$ is the quasiparticle velocity. The fermion mass current density is then simply

$$m\mathbf{j} = 2m \int (d\mathbf{p})\, \nabla_{\mathbf{p}}\epsilon_{\mathbf{p}}\, n_{\mathbf{p}}, \tag{1.38}$$

where we used $\mathbf{v}_{\mathbf{p}} = \nabla_{\mathbf{p}}\epsilon_{\mathbf{p}}$. Eq.(1.38) represents the momentum transported by the fermions, which is the same as the momentum transported by the quasiparticles. Consequently,

$$\int (d\mathbf{p})\, \mathbf{p}\, n_{\mathbf{p}} = m \int (d\mathbf{p})\, \nabla_{\mathbf{p}}\epsilon_{\mathbf{p}}\, n_{\mathbf{p}}. \tag{1.39}$$

Next we vary Eq.(1.39) with respect to $n_{\mathbf{p}}$ keeping in mind that the quasiparticle energy $\epsilon_{\mathbf{p}}$ is itself a functional of $n_{\mathbf{p}}$. Assume for the moment that there is no magnetic field, so $\epsilon_{\mathbf{p}}$ and $n_{\mathbf{p}}$ do not depend on the spin. Then both $\epsilon_{\mathbf{p}}$ and $n_{\mathbf{p}}$ are diagonal in the spin indices, and have the same value for the two spin eigenstates. We can therefore drop the spin index from $\epsilon_{\mathbf{p}}$ and $n_{\mathbf{p}}$. Eq.(1.26) tells us that the variation of $\epsilon_{\mathbf{p}}$ in the present condition is

$$\delta\epsilon_{\mathbf{p}} = \frac{1}{2}\mathrm{Tr}_\sigma \mathrm{Tr}_{\sigma'} \int (d\mathbf{p}')\, f(\mathbf{p},\boldsymbol{\sigma};\mathbf{p}',\boldsymbol{\sigma}')\delta n_{\mathbf{p}'}(\boldsymbol{\sigma}'),$$

where the trace over the two spin indices represents the sum $\sum_\alpha \sum_\beta f_{\alpha\alpha,\beta\beta}$. Varying Eq.(1.39) we get

$$\begin{aligned}\frac{1}{m}\int (d\mathbf{p})\mathbf{p}\delta n_{\mathbf{p}} = &\int (d\mathbf{p})\nabla_{\mathbf{p}}\epsilon_{\mathbf{p}}\delta n_{\mathbf{p}} \\ &+ \frac{1}{2}\mathrm{Tr}_\sigma \mathrm{Tr}_{\sigma'} \int (d\mathbf{p})(d\mathbf{p}')\, n_{\mathbf{p}}\nabla_{\mathbf{p}} f(\mathbf{p},\boldsymbol{\sigma};\mathbf{p}',\boldsymbol{\sigma}')\delta n_{\mathbf{p}'}.\end{aligned}$$

We now integrate by parts (with respect to $\mathbf{p}$) the second term on the right hand side of the above equation:

$$\begin{aligned}\frac{1}{m}\int (d\mathbf{p})\mathbf{p}\delta n_{\mathbf{p}} = &\int (d\mathbf{p})\nabla_{\mathbf{p}}\epsilon_{\mathbf{p}}\delta n_{\mathbf{p}} \\ &- \frac{1}{2}\mathrm{Tr}_\sigma \mathrm{Tr}_{\sigma'} \int (d\mathbf{p})(d\mathbf{p}')\, (\nabla_{\mathbf{p}} n_{\mathbf{p}})\, f(\mathbf{p},\boldsymbol{\sigma};\mathbf{p}',\boldsymbol{\sigma}')\delta n_{\mathbf{p}'}.\end{aligned}$$

Finally we interchange the variables $\mathbf{p},\boldsymbol{\sigma}$ and $\mathbf{p}',\boldsymbol{\sigma}'$ in the second term on the right hand side:

$$\begin{aligned}\frac{1}{m}\int (d\mathbf{p})\mathbf{p}\delta n_{\mathbf{p}} = &\int (d\mathbf{p})\nabla_{\mathbf{p}}\epsilon_{\mathbf{p}}\delta n_{\mathbf{p}} \\ &- \frac{1}{2}\mathrm{Tr}_\sigma \mathrm{Tr}_{\sigma'} \int (d\mathbf{p})(d\mathbf{p}')\, (\nabla_{\mathbf{p}'} n_{\mathbf{p}'})\, f(\mathbf{p}',\boldsymbol{\sigma}';\mathbf{p},\boldsymbol{\sigma})\delta n_{\mathbf{p}}.\end{aligned}$$

Since $\delta n_{\mathbf{p}}$ is arbitrary, it follows that

$$\frac{\mathbf{p}}{m} = \nabla_{\mathbf{p}}\epsilon_{\mathbf{p}} - \frac{1}{2}\mathrm{Tr}_\sigma \mathrm{Tr}_{\sigma'} \int (d\mathbf{p}')\, (\nabla_{\mathbf{p}'} n_{\mathbf{p}'})\, f(\mathbf{p}',\boldsymbol{\sigma}';\mathbf{p},\boldsymbol{\sigma}). \tag{1.40}$$

There is a little more to go before we get our final expression for the effective mass m^*. Eq.(1.25) tells us that the $T = 0$ equilibrium distribution function is a step function,

$$n_{\mathbf{p}} = \Theta(\mu(0) - \epsilon_{\mathbf{p}}), \tag{1.41}$$

and if we use the linearized dispersion $\epsilon_{\mathbf{p}} - \mu(0) = v(p - p_F)$ of Eq.(1.34), the gradient of $n_{\mathbf{p}}$ takes a simple form,

$$\nabla_{\mathbf{p}'} n_{\mathbf{p}'} = -\hat{\mathbf{p}}'\delta(p' - p_F). \tag{1.42}$$

Using this in Eq.(1.40) along with the definition of the effective mass in Eq.(1.35) (i.e., $v = p_F/m^* = (1/\hbar)|\nabla_{\mathbf{p}}\epsilon_{\mathbf{p}}|_{p=p_F}$) we arrive at the following relation for $\mathbf{p}$ lying on the Fermi surface $p = p_F$:

$$\frac{p_F\hat{\mathbf{p}}}{m} = \frac{p_F\hat{\mathbf{p}}}{m^*} + \frac{1}{2(2\pi\hbar)^3}\mathrm{Tr}_\sigma \mathrm{Tr}_{\sigma'} \int p'^2 dp' d\Omega_{\mathbf{p}'}\hat{\mathbf{p}}'\delta(p' - p_F)\, f(\mathbf{p}', \boldsymbol{\sigma}'; \mathbf{p}, \boldsymbol{\sigma}). \tag{1.43}$$

Here $d\Omega_{\mathbf{p}'}$ represents the solid angle subtended at the origin by the volume element $d\mathbf{p}'$. In an isotropic system, we discussed earlier that $f(\mathbf{p}', \boldsymbol{\sigma}'; \mathbf{p}, \boldsymbol{\sigma})$ will depend not on the individual directions of $\mathbf{p}$ and $\mathbf{p}'$ but on their mutual angular separation χ. If $\mathbf{p}$ and $\mathbf{p}'$ both lie on the Fermi surface as is the case in Eq.(1.43), then $f(\mathbf{p}', \boldsymbol{\sigma}'; \mathbf{p}, \boldsymbol{\sigma})$ is a function only of the angle χ. Taking the scalar product with $\hat{\mathbf{p}}$ on both sides of Eq.(1.43) and rearranging, we get an expression for the effective mass m^* in terms of the Fermi liquid parameter f :

$$\begin{aligned} \frac{1}{m^*} &= \frac{1}{m} - \frac{p_F}{2(2\pi\hbar)^3}\mathrm{Tr}_\sigma \mathrm{Tr}_{\sigma'} \int d\Omega \cos\chi\, f(\chi; \boldsymbol{\sigma}, \boldsymbol{\sigma}'). \\ &= \frac{1}{m} - \frac{4\pi p_F}{(2\pi\hbar)^3}\int_0^\pi d\chi\, \sin\chi\, \cos\chi\, f^s(\chi). \end{aligned} \tag{1.44}$$

The above expression can be presented in a more appealing form by introducing dimensionless Fermi liquid parameter F, where F is related to the original parameter f through a rescaling by the density of states $\nu(0)$ at the $T = 0$ chemical potential, $\mu(0)$:

$$F = \nu(0) f = \frac{m^* p_F}{\pi^2\hbar^3} f. \tag{1.45}$$

Eq.(1.44) then takes the form

$$\begin{aligned} \frac{1}{m^*} &= \frac{1}{m} - \frac{1}{2m^*}\int_0^\pi d\chi\, \sin\chi\, \cos\chi\, F^s(\chi) \\ &= \frac{1}{m} - \frac{1}{2m^*}\sum_{l=0}^{\infty}\int_0^\pi d\chi\, \sin\chi\, \cos\chi\, F_l^s P_l(\cos\chi) \\ &= \frac{1}{m} - \frac{F_1^s}{3m^*}. \end{aligned}$$

Upon rearranging,

$$\frac{m^*}{m} = 1 + \frac{F_1^s}{3}. \tag{1.46}$$

The specific heat can be estimated in the same way as for the noninteracting case. The only difference is that the effective mass appears in Eq.(1.12) instead of the bare mass m :

$$C_V = \gamma T = \frac{m^* p_F k_B^2}{3\hbar^3} T. \tag{1.47}$$

1.3.2 Compressibility and speed of sound

The compressibility κ is defined as

$$\frac{1}{\kappa} = -V\frac{\partial P}{\partial V}, \tag{1.48}$$

where P is the pressure. We know that the speed of sound, s, in a medium depends on the compressibility and the mass density $\rho = mn$:

$$s^2 = \frac{1}{\kappa\rho}. \tag{1.49}$$

Such sound waves involve a collective motion of a large number of particles – matter is essentially treated as a continuum. This is the so called hydrodynamic limit. In a Fermi liquid, we have seen that the relaxation rate τ^{-1} (due to collisions) evolves as T^2. For a given frequency, ω, there is a range of frequencies where $\omega\tau \ll 1$ is satisfied. In this regime, the collisions distribute the quasiparticle momenta and energy uniformly over a period of oscillation, and the sound propagation is a hydrodynamic process.

Longitudinal sound waves in, say, an ideal gas, are usually regarded as a succession of *adiabatic* compressions and rarefactions - adiabatic because the frequency of the waves is usually larger than the rate at which the gas is able to come to *thermal* equilibrium with the surroundings during the course of a compression-rarefaction cycle. So in an ideal gas we are usually interested in the adiabatic (i.e., constant entropy) compressibility. However near $T = 0$, there is not much difference between an adiabatic and an isothermal process. Indeed, at absolute zero, the third law of thermodynamics tells us that all processes involve zero entropy change. The isothermal compressibility can be easily obtained in the following manner. From the identities

$$d\Omega = -SdT - Nd\mu,$$

and

$$\Omega = -PV,$$

we have

$$N = -\left(\frac{\partial\Omega}{\partial\mu}\right)_{V,T} = V\left(\frac{\partial P}{\partial\mu}\right)_{V,T} = V\left(\frac{\partial P}{\partial N}\right)_{V,T}\left(\frac{\partial N}{\partial\mu}\right)_{V,T}. \tag{1.50}$$

Now we use $(\partial P/\partial N)_V = -(\partial P/\partial V)_N(\partial V/\partial N)_P = -(V/N)(\partial P/\partial V)_N$ in Eq.(1.50) along with the relation for the speed of sound, Eq.(1.49), to get

$$\frac{N}{m}\left(\frac{\partial\mu}{\partial N}\right)_{V,T} = -\frac{V^2}{Nm}\left(\frac{\partial P}{\partial V}\right)_{N,T} = \frac{1}{\kappa\rho} = s^2. \tag{1.51}$$

Thus to calculate the speed of sound in a Fermi liquid, we need to know

$(\partial\mu/\partial N)_{V,T}$. Consider adding a small number, dN, of fermions to the system with fixed volume V. The Fermi energy will change by an amount $d\mu$. This change has two contributions. First, just as in the noninteracting case, adding fermions to the system will increase the Fermi momentum, and therefore, the Fermi energy. Second, the finite interaction energy due to the added extra fermions will also change the chemical potential. Combining these two contributions we can write,

$$d\mu = v_{\mathbf{p}}\, dp_F + \frac{1}{2}\mathrm{Tr}_{\boldsymbol{\sigma}}\mathrm{Tr}_{\boldsymbol{\sigma}'} \int (d\mathbf{p}') f(\mathbf{p},\boldsymbol{\sigma};\mathbf{p}',\boldsymbol{\sigma}')\delta n_{\mathbf{p}'}. \tag{1.52}$$

At $T = 0$, the change in the distribution function is

$$\delta n_{\mathbf{p}} \approx \delta(\epsilon^{(0)}(\mathbf{p}) - \mu) v_{\mathbf{p}} dp_F, \tag{1.53}$$

and therefore,

$$dN = 2V \int (d\mathbf{p})\delta n_{\mathbf{p}} = 2V \int (d\mathbf{p})\delta(\epsilon^{(0)}(\mathbf{p}) - \mu) v_{\mathbf{p}} dp_F. \tag{1.54}$$

The speed of sound is given by

$$\frac{N}{ms^2} = \frac{\partial N}{\partial \mu} = 2V \int (d\mathbf{p})\delta(\epsilon^{(0)}(\mathbf{p}) - \mu) v_{\mathbf{p}} \frac{dp_F}{d\mu}. \tag{1.55}$$

Upon substituting Eq.(1.53) in Eq.(1.52), we obtain

$$1 = v_{\mathbf{p}} \frac{dp_F}{d\mu} + \frac{1}{2}\mathrm{Tr}_{\boldsymbol{\sigma}}\mathrm{Tr}_{\boldsymbol{\sigma}'} \int (d\mathbf{p}') f(\mathbf{p},\boldsymbol{\sigma};\mathbf{p}',\boldsymbol{\sigma}')\delta(\epsilon^{(0)}(\mathbf{p}') - \mu) v_{\mathbf{p}'} \frac{dp'_F}{d\mu}. \tag{1.56}$$

Now we multiply both sides of Eq.(1.56) by $2\delta(\epsilon^{(0)}(\mathbf{p}) - \mu)$ and integrate over $\mathbf{p}$. The left hand side, $2\int(d\mathbf{p})\delta(\epsilon^{(0)}(\mathbf{p}) - \mu)$ is, by definition, simply equal to the density of states $\nu(0) = m^* p_F/(\pi^2\hbar^3)$ at the chemical potential $\mu(0)$. The second term on the right hand side of Eq.(1.56) will look like

$$\mathrm{Tr}_{\boldsymbol{\sigma}}\mathrm{Tr}_{\boldsymbol{\sigma}'} \int (d\mathbf{p}')(d\mathbf{p}) f(\mathbf{p},\boldsymbol{\sigma};\mathbf{p}',\boldsymbol{\sigma}')\delta(\epsilon^{(0)}(\mathbf{p}') - \mu)\delta(\epsilon^{(0)}(\mathbf{p}) - \mu) v_{\mathbf{p}'} \frac{dp'_F}{d\mu}.$$

If we interchange $\mathbf{p},\boldsymbol{\sigma} \leftrightarrow \mathbf{p}',\boldsymbol{\sigma}'$, then the above term looks like

$$\mathrm{Tr}_{\boldsymbol{\sigma}}\mathrm{Tr}_{\boldsymbol{\sigma}'} \int (d\mathbf{p}')(d\mathbf{p}) f(\mathbf{p}',\boldsymbol{\sigma}';\mathbf{p},\boldsymbol{\sigma})\delta(\epsilon^{(0)}(\mathbf{p}') - \mu)\delta(\epsilon^{(0)}(\mathbf{p}) - \mu) v_{\mathbf{p}} \frac{dp_F}{d\mu}.$$

Gathering these results into Eq.(1.56), we have

$$\nu(0) = 2\int (d\mathbf{p})\delta(\epsilon^{(0)}(\mathbf{p}) - \mu) v_{\mathbf{p}} \frac{dp_F}{d\mu} \times \tag{1.57}$$

$$\times \left[1 + \frac{1}{2}\mathrm{Tr}_{\boldsymbol{\sigma}}\mathrm{Tr}_{\boldsymbol{\sigma}'} \int (d\mathbf{p}') f(\mathbf{p}',\boldsymbol{\sigma}';\mathbf{p},\boldsymbol{\sigma})\delta(\epsilon^{(0)}(\mathbf{p}') - \mu)\right]. \tag{1.58}$$

Now we expand f in a series of Legendre polynomials as in Eq.(1.33). Then using the relation $\nu(0)f = F$ that we introduced in Eq.(1.45), Eq.(1.58) takes the simple form

$$\frac{\nu(0)V}{1+F_0^s} = \left(\frac{\partial N}{\partial \mu}\right) = \frac{N}{ms^2}.$$

The final expression for the speed of sound s is

$$s^2 = \frac{N(1+F_0^s)}{mV\nu(0)} = \frac{p_F^2}{3mm^*}(1+F_0^s). \tag{1.59}$$

If $(1+F_0^s) < 0$, then the speed of sound given by Eq.(1.59) will not remain real. Such a situation signifies an instability. Another way to see this is to look at the compressibility, $\kappa = 1/(s^2\rho)$. If the compressibility diverges, then the system becomes unstable because in that case a small external pressure will give rise to very large volume strain.

1.3.3 Pauli susceptibility

For a free fermion, the extra energy in an external magnetic field $\mathbf{B}_{\text{ext}}$ will be $-\mu_B \boldsymbol{\sigma} \cdot \mathbf{B}_{\text{ext}}$. In a Fermi liquid, there is an extra contribution due to electron interactions because the distribution function also gets affected by the magnetic field. The change in quasiparticle energy due to a small $\mathbf{B}_{\text{ext}}$ can be written as

$$\delta\epsilon_{\mathbf{p}}(\boldsymbol{\sigma}) = -\mu_B \boldsymbol{\sigma} \cdot \mathbf{B}_{\text{ext}} + \text{Tr}_{\boldsymbol{\sigma}'} \int (d\mathbf{p}') f(\mathbf{p}, \boldsymbol{\sigma}; \mathbf{p}', \boldsymbol{\sigma}') \delta n_{\mathbf{p}'}(\boldsymbol{\sigma}'). \tag{1.60}$$

The magnetic field causes a change δn in the distribution function near the Fermi surface, which is related to $\delta\epsilon$ through

$$\delta n_{\mathbf{p}}(\boldsymbol{\sigma}) = -\delta(\epsilon_{\mathbf{p}} - \mu)\delta\epsilon_{\mathbf{p}}(\boldsymbol{\sigma}).$$

Thus $\delta\epsilon$ satisfies

$$\delta\epsilon_{\mathbf{p}}(\boldsymbol{\sigma}) = -\mu_B \boldsymbol{\sigma} \cdot \mathbf{B}_{\text{ext}} - \text{Tr}_{\boldsymbol{\sigma}'} \int (d\mathbf{p}') f(\mathbf{p}, \boldsymbol{\sigma}; \mathbf{p}', \boldsymbol{\sigma}') \delta(\epsilon_{\mathbf{p}'} - \mu) \delta\epsilon_{\mathbf{p}'}(\boldsymbol{\sigma}'). \tag{1.61}$$

We are looking for a solution of the form

$$\delta\epsilon_{\mathbf{p}}(\boldsymbol{\sigma}) = -\frac{g_{\text{eff}}}{2} \mu_B \boldsymbol{\sigma} \cdot \mathbf{B}_{\text{ext}} \tag{1.62}$$

on the Fermi surface $|\mathbf{p}| = p_F$. With both $\mathbf{p}$ and $\mathbf{p}'$ lying on the Fermi surface, f depends, as usual, only on the angle χ between $\mathbf{p}$ and $\mathbf{p}'$. Multiplying both sides of Eq.(1.61) by $\delta(\epsilon_{\mathbf{p}} - \mu)$ and integrating over $\mathbf{p}$ we have

$$-\frac{(g_{\text{eff}} - 2)}{g_{\text{eff}}} \boldsymbol{\sigma} \cdot \mathbf{B}_{\text{ext}} \int (d\mathbf{p}) \delta(\epsilon_{\mathbf{p}} - \mu)$$
$$= \text{Tr}_{\boldsymbol{\sigma}'} \int (d\mathbf{p}')(d\mathbf{p}) f(\mathbf{p}, \boldsymbol{\sigma}; \mathbf{p}', \boldsymbol{\sigma}') \delta(\epsilon_{\mathbf{p}} - \mu) \delta(\epsilon_{\mathbf{p}'} - \mu) \boldsymbol{\sigma}' \cdot \mathbf{B}_{\text{ext}},$$

or, by using $\nu(0) = 2\int(d\mathbf{p})\delta(\epsilon_{\mathbf{p}} - \mu)$,

$$-\frac{(g_{\text{eff}} - 2)}{g_{\text{eff}}}\boldsymbol{\sigma}\cdot\mathbf{B}_{\text{ext}} = \frac{\nu(0)}{4}\text{Tr}_{\boldsymbol{\sigma}'}\int_0^{\pi} d\chi\,\sin\chi\, f(\chi;\boldsymbol{\sigma},\boldsymbol{\sigma}')\boldsymbol{\sigma}'\cdot\mathbf{B}_{\text{ext}}. \quad (1.63)$$

Now $f(\boldsymbol{\sigma},\boldsymbol{\sigma}') = f^{(s)} + (\boldsymbol{\sigma}\cdot\boldsymbol{\sigma}')f^{(a)}$. We now take the trace over $\boldsymbol{\sigma}'$ in Eq.(1.63) making use of the fact that

$$\text{Tr}\sigma = 0, \text{ and,}$$
$$\text{Tr}_{\boldsymbol{\sigma}'}[(\boldsymbol{\sigma}\cdot\boldsymbol{\sigma}')\boldsymbol{\sigma}'] = 2\boldsymbol{\sigma},$$

we obtain

$$g_{\text{eff}} = \frac{2}{1 + F_0^a}. \quad (1.64)$$

We can now determine the susceptibility, χ, which is related to the magnetization $\mathbf{M}$ through

$$\mathbf{M} = \chi\mathbf{B}_{\text{ext}} = \mu_B\text{Tr}_{\boldsymbol{\sigma}}\int(d\mathbf{p})\boldsymbol{\sigma}\delta n_{\mathbf{p}}(\boldsymbol{\sigma}) = -\mu_B\text{Tr}_{\boldsymbol{\sigma}}\int(d\mathbf{p})\boldsymbol{\sigma}\delta(\epsilon_{\mathbf{p}} - \mu)\delta\epsilon_{\mathbf{p}}(\boldsymbol{\sigma}).$$

Using Eq.(1.62) in the above relation, we have

$$\chi\mathbf{B}_{\text{ext}} = \frac{\nu(0)g_{\text{eff}}\mu_B^2}{4}\text{Tr}[\boldsymbol{\sigma}(\boldsymbol{\sigma}\cdot\mathbf{B}_{\text{ext}})], \text{ or,}$$
$$\chi = \frac{\nu(0)g_{\text{eff}}\mu_B^2}{2} = \frac{\mu_B^2 p_F m^*}{\pi^2\hbar^3(1 + F_0^a)}. \quad (1.65)$$

The susceptibility χ is the first quantity we have encountered so far that depends on the (spin) *antisymmetric* Fermi liquid parameter F^a. Both the effective mass and the compressibility depended on the symmetric parameter F^s. Unlike the noninteracting case, the ratio of the coefficient of linear$-T$ specific heat, γ, to the Pauli susceptibility, χ, is no longer a universal constant in case of the Fermi liquid.

Observe that the susceptibility diverges as $(1 + F_0^a) \to 0$. When that happens, the system becomes unstable against a ferromagnetic transition.

1.4 Landau postulates and Bethe ansatz integrable systems

We have seen that the Landau Fermi liquid theory rests on some key postulates. The first one is that eigenstates of the interacting system are in one-to-one correspondence with those of the noninteracting counterpart. Second, the elementary excitations of the interacting system may be labeled by the momenta just as in the noninteracting limit. Apart from the momentum, the eigenstates in general carry other labels such as the spin. The ground state is constructed by filling the fermions in the energy levels beginning with the one with the smallest momentum. Finally, the energy of an excitation corresponding to a given momentum depends on the occupation of states with different momenta. This, in fact, encapsulates the effects of interparticle interaction.

These postulates nevertheless do not uniquely define a Landau Fermi liquid. A counterexample is provided by integrable fermionic systems solvable by the Bethe ansatz. Consider for example a model of M spinless, repulsively interacting fermions on a one-dimensional ring with $N(> 2M)$ sites:

$$H = -\sum_{j=1}^{N}[c_j^\dagger c_{j+1} + \text{h.c.}] + U\sum_j n_j n_{j+1}. \tag{1.66}$$

Here $U(> 0)$ is the repulsive interaction of fermions on neighboring sites. We assume N is even and M is odd. The M-particle eigenstates are labeled by the particle momenta k_j that are determined by the Bethe equations[6],

$$Nk_j = 2\pi I_j + \sum_{l=1}^{M}\theta(k_j, k_l), \tag{1.67}$$

$$\theta(k_j, k_l) = 2\tan^{-1}\left(\frac{U\sin[\frac{1}{2}(k_j - k_l)]}{\cos[\frac{1}{2}(k_j + k_l)] - U\cos[\frac{1}{2}(k_j - k_l)]}\right). \tag{1.68}$$

Here I_j are integers defining the state. The momenta k_j smoothly evolve from the noninteracting values $k_j = 2\pi I_j/N$ as the interaction U is increased from zero. Without going into details, we state that this is not a Landau Fermi liquid. For such models, the single-particle Green function does not have the simple pole structure expected of a Landau quasiparticle. We will study the analytic properties of the single-particle Green functions in Chapter 9.

[6]The model in Eq. (1.66) is equivalent to the XXZ spin-1/2 antiferromagnetic chain and has been solved, for example, in Bill Sutherland, *Beautiful Models* (World Scientific, Singapore, 2004).

Chapter 2

Pomeranchuk Stability Conditions

In Sec. 1.3 we found that the stability of the Fermi liquid against a density instability required the condition $(1 + F_0^s) > 0$ to be satisfied, and the stability condition against a ferromagnetic transition was similarly shown to be $(1 + F_0^a) > 0$. We showed that the density and ferromagnetic instabilities were associated with the divergence of the respective response functions, namely, the compressibility ($\kappa \propto 1/(1+F_0^s)$) and the Pauli spin susceptibility ($\chi \propto 1/(1 + F_0^a)$). These two examples happen to be special cases of more general conditions for the stability of the Fermi liquid against small arbitrary perturbations $\delta n_{\mathbf{p}}(\boldsymbol{\sigma})$. They are

$$
\begin{aligned}
1 + \frac{F_l^s}{2l+1} &> 0, \\
1 + \frac{F_l^a}{2l+1} &> 0.
\end{aligned}
\tag{2.1}
$$

These conditions were derived by Pomeranchuk in 1958, soon after Landau published his landmark Fermi liquid papers. When these conditions are not satisfied, the resulting instability is known as a Pomeranchuk instability.

2.1 Fermi surface deformation and excitation free energy

The total excitation free energy δF consists of a noninteracting quasiparticle part, and an "f" part originating in the interaction of the fermions:

$$
\delta F = \mathrm{Tr}_{\boldsymbol{\sigma}} \int (d\mathbf{p})(\epsilon_{\mathbf{p}}^{(0)}(\boldsymbol{\sigma}) - \mu)\delta n_{\mathbf{p}}(\boldsymbol{\sigma}) \tag{2.2}
$$

$$
+ \frac{1}{2}\mathrm{Tr}_{\boldsymbol{\sigma}}\mathrm{Tr}_{\boldsymbol{\sigma}'} \int (d\mathbf{p})(d\mathbf{p}') f(\mathbf{p}, \boldsymbol{\sigma}; \mathbf{p}', \boldsymbol{\sigma}')\delta n_{\mathbf{p}}(\boldsymbol{\sigma})\delta n_{\mathbf{p}'}(\boldsymbol{\sigma}'), \tag{2.3}
$$

where $\epsilon_{\mathbf{p}}^{(0)} - \mu = v_{\mathbf{p}}(p - p_F) = v_F(p - p_F)$, and $v_F = p_F/m^*$. Stability will exist if $\delta F > 0$ for small and arbitrary $\delta n_{\mathbf{p}}(\boldsymbol{\sigma})$. The $\delta n_{\mathbf{p}}(\boldsymbol{\sigma})$ will distort the

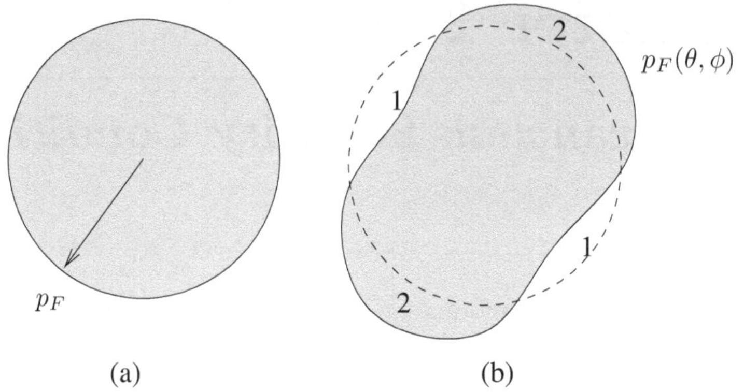

FIGURE 2.1: Schematic of the distribution function $n_{\mathbf{p}}$ (a) in the unperturbed condition, and (b) following a perturbation $\delta n_{\mathbf{p}}$. The shaded areas represent filled states. In regions marked as 1, $\delta n = -1$ and in the regions marked as 2, $\delta n = +1$ with respect to the unperturbed distribution. $p_F(\theta, \phi)$ is the new position of the Fermi surface.

Fermi surface so that some of the previously unoccupied states will now get occupied, and vice versa. Figure 2.1 shows an example of a distorted Fermi surface following a small perturbation of the distribution function.

Let us express the small perturbation of the Fermi momentum in Figure 2.1 as an expansion in spherical harmonics:

$$p_F(\theta, \phi) = p_F + \sum_{l,m} \Phi_{lm} Y_{lm}(\theta, \phi). \tag{2.4}$$

The spherical harmonics Y_{lm} can be expressed in terms of the associated Legendre polynomials $P_l^m(\cos\theta)$:

$$Y_{lm}(\theta, \phi) = C_{lm} e^{im\phi} P_l^m(\cos\theta). \tag{2.5}$$

If the Y_{lm} are to satisfy the orthogonality condition

$$\int d\Omega Y_{lm}(\theta, \phi) Y^*_{l'm'}(\theta, \phi) = \delta_{ll'} \delta_{mm'}, \tag{2.6}$$

the normalization constants C_{lm} should be

$$C_{lm} = \sqrt{\frac{(2l+1)}{4\pi} \frac{(l-m)!}{(l+m)!}}.$$

Here $d\Omega = d\theta\, d\phi\, \sin\theta$ as usual.

Consider first the situation where the Φ_{lm} are independent of the spin.

The first term in the expression for the free energy δF shown in Eq.(2.3) is then

$$\frac{2v_F p_F^2}{(2\pi\hbar)^3}\int d\Omega \int_{p_F}^{p_F(\theta,\phi)} dp\,(p-p_F) = \frac{v_F p_F^2}{(2\pi\hbar)^3}\int d\Omega\,(p_F(\theta,\phi)-p_F)^2. \quad (2.7)$$

In order to complete the integral on the right hand side of Eq.(2.7) (and more), we need to recapitulate a few relevant relations involving the spherical harmonics Y_{lm}. First, a relation for the complex conjugate of a spherical harmonic:

$$Y^*_{lm}(\theta,\phi) = (-1)^m Y_{l,-m}(\theta,\phi). \quad (2.8)$$

Second, we require the *addition theorem* for spherical harmonics. Let χ be the angle between the directions (θ,ϕ) and (θ',ϕ'). The cosine of χ is given by

$$\cos\chi = \cos\theta\,\cos\theta' + \sin\theta\,\sin\theta'\cos(\phi-\phi').$$

This can be checked by simply taking the dot product of the unit vector $(\sin\theta\,\cos\phi, \sin\theta\,\sin\phi, \cos\theta)$ with the corresponding one for (θ',ϕ'). The addition theorem states that

$$P_l(\cos\chi) = \frac{4\pi}{2l+1}\sum_{m=-l}^{l} Y^*_{lm}(\theta',\phi')Y_{lm}(\theta,\phi). \quad (2.9)$$

The term on the right hand side of Eq.(2.7) is

$$\frac{v_F p_F^2}{(2\pi\hbar)^3}\int d\Omega\,(p_F(\theta,\phi)-p_F)^2$$
$$= \frac{v_F p_F^2}{(2\pi\hbar)^3}\sum_{l_1,m_1}\sum_{l_2,m_2}\Phi_{l_1m_1}\Phi_{l_2m_2}\int d\Omega Y_{l_1m_1}(\theta,\phi)Y_{l_2m_2}(\theta,\phi).$$

We can write $Y_{l_2m_2} = (-1)^{m_2}Y_{l_2,-m_2}$ making use of the relation for complex conjugation, Eq.(2.8). Now we complete the angular integration using the orthogonality relation, Eq.(2.6). We get

$$\frac{v_F p_F^2}{(2\pi\hbar)^3}\int d\Omega\,(p_F(\theta,\phi)-p_F)^2 = \frac{v_F p_F^2}{(2\pi\hbar)^3}\sum_{l_1,m_1}\Phi_{l_1m_1}\Phi_{l_1,-m_1}(-1)^{m_1}. \quad (2.10)$$

We can simplify this further by making use of the fact that $\Phi(\theta,\phi) = \sum_{lm} Y_{lm}\Phi_{lm}$ is a real function; so $\Phi^*(\theta,\phi) = \Phi(\theta,\phi)$. Now

$$\Phi^*(\theta,\phi) = \sum_{l,m} Y^*_{lm}\Phi^*_{lm} = \sum_{l,m} Y_{l,-m}(-1)^m\Phi^*_{lm} \equiv \sum_{l,m} Y_{lm}(-1)^{-m}\Phi^*_{l,-m}.$$

Thus we obtain the relation

$$\Phi_{lm} = (-1)^{-m}\Phi^*_{l,-m}, \text{ or,}$$
$$\Phi^*_{lm} = (-1)^{-m}\Phi_{l,-m}. \quad (2.11)$$

Eq.(2.11) enables us to write Eq.(2.10) as

$$\frac{v_F p_F^2}{(2\pi\hbar)^3}\int d\Omega\,(p_F(\theta,\phi)-p_F)^2=\frac{v_F p_F^2}{(2\pi\hbar)^3}\sum_{l,m}|\Phi_{lm}|^2. \tag{2.12}$$

This completes our calculation of the first term on the right hand side of Eq.(2.3).

Consider now the second term on the sight hand side of Eq.(2.3) representing the interaction energy. This term reduces to

$$\frac{2p_F^4}{(2\pi\hbar)^6}\int d\Omega\int d\Omega' f^s(\chi)[p_F(\theta,\phi)-p_F][p_F(\theta',\phi')-p_F],$$

where χ, as we have said, is the angle between the directions (θ,ϕ) and (θ',ϕ'). We expand $f(\chi)$ in a series of Legendre polynomials $P_l(\cos\chi)$, and use the addition theorem of Eq.(2.9) to resolve the Legendre polynomial into a sum over the product of two spherical harmonics. Thus the interaction energy takes the form

$$\frac{2p_F^4}{(2\pi\hbar)^6}\sum_{lm}\left(\frac{4\pi}{2l+1}\right)f_l^s\int d\Omega\int d\Omega' Y_{lm}^*(\theta',\phi')Y_{lm}(\theta,\phi)\times$$
$$\times\sum_{l_1,m_1}\sum_{l_2,m_2}\Phi_{l_1m_1}\Phi_{l_2m_2}Y_{l_1m_1}(\theta,\phi)Y_{l_2m_2}(\theta',\phi').$$

The integration over Ω' is straightforward. Using the relation for complex conjugation, Eq.(2.8), we can also perform the integration over Ω as we did in the case of the first term of Eq.(2.3). After that we use the relation we obtained in Eq.(2.11) to obtain the following final form for the interaction energy,

$$\frac{2p_F^4}{(2\pi\hbar)^6}\sum_{l,m}\left(\frac{4\pi}{2l+1}\right)f_l^s|\Phi_{lm}|^2.$$

Combining this with Eq.(2.12), we get the free energy functional,

$$\delta F=\frac{v_F p_F^2}{(2\pi\hbar)^3}\sum_{l,m}|\Phi_{lm}|^2+\frac{2p_F^4}{(2\pi\hbar)^6}\sum_{l,m}\left(\frac{4\pi}{2l+1}\right)f_l^s|\Phi_{lm}|^2. \tag{2.13}$$

2.1.1 Pomeranchuk stability conditions

The requirement for stability is that the free energy δF be positive. Since $|\Phi_{lm}|^2$ is always positive, we must satisfy

$$\frac{v_F p_F^2}{(2\pi\hbar)^3}+\frac{2p_F^4}{(2\pi\hbar)^6}\left(\frac{4\pi}{2l+1}\right)f_l^s>0,\text{ or,}$$
$$1+\frac{m^* p_F}{\pi^2\hbar^3}\frac{f_l^s}{2l+1}=1+\frac{F_l^s}{2l+1}>0. \tag{2.14}$$

This is the first of the two Pomeranchuk stability conditions we mentioned at the beginning of the present chapter. We briefly sketch the proof of the second Pomeranchuk stability condition.

Suppose the $\delta n_{\mathbf{p}}(\boldsymbol{\sigma})$ do depend on $\boldsymbol{\sigma}$. Then the resulting Fermi momentum will also depend on $\boldsymbol{\sigma}$. We introduce a spin-dependent perturbation of p_F :

$$p_F(\theta, \phi; \boldsymbol{\sigma}) = p_F + \sum_{l,m} \boldsymbol{\sigma} \Phi_{lm} Y_{lm}(\theta, \phi). \tag{2.15}$$

The excitation energy, Eq.(2.3), can be evaluated as before, and we get

$$\delta F = \frac{v_F p_F^2}{(2\pi\hbar)^3} \sum_{l,m} |\Phi_{lm}|^2 + \frac{2p_F^4}{(2\pi\hbar)^6} \sum_{l,m} \left(\frac{4\pi}{2l+1}\right) f_l^a |\Phi_{lm}|^2. \tag{2.16}$$

The stability condition $\delta F > 0$ leads to

$$1 + \frac{F_l^a}{2l+1} > 0, \tag{2.17}$$

which is the second Pomeranchuk condition for stability. If for a certain l, $F_l^{s(a)} \leq -(2l+1)$, then the system becomes unstable against a phase transition in that channel. As the coefficient of $|\Phi_{lm}|^2$ becomes negative, higher order terms (e.g. $|\Phi_{lm}|^4$) in the free energy will eventually stabilize the Fermi liquid at a finite mean value of $|\Phi_{lm}|^2$, or in other words, a finite distortion of the Fermi surface.

Finally a word for the special case $l = 1$. It was shown by Leggett[1] in 1968 that the inequality

$$F_1^s \geq F_1^a \tag{2.18}$$

is obeyed in a Fermi liquid. Equality is possible in certain pathological cases where charge and spin currents are both conserved.

2.2 Stoner ferromagnetism

The special case of a Pomeranchuk instability in the $l = 0$ spin antisymmetric channel was studied as far back as 1938 by Stoner, who also proposed a physical mechanism for this ferromagnetic instability. Exchange interaction of the electrons reduces the energy cost of the deformation of the Fermi surface associated with unequal populations of electrons with opposite spin polarizations. For a sufficiently large exchange interaction, the energy cost of this

[1] A. J. Leggett, *Annals of Physics* **46**, 76 (1968).

deformation turns negative, leading to appearance of a nonzero spin polarization.

Consider a simple model of electrons with (noninteracting) energy spectrum $\epsilon_{\mathbf{p}}^{(0)}$ and an exchange interaction of strength I :

$$F = \sum_{\sigma} \int (d\mathbf{p})(\epsilon_{\mathbf{p}}^{(0)} - \mu) n_{\mathbf{p}}(\sigma) + I \sum_{\sigma} \int (d\mathbf{p})(d\mathbf{p}') n_{\mathbf{p}}(\sigma) n_{\mathbf{p}'}(\bar{\sigma}), \qquad (2.19)$$

where $\bar{\sigma}$ refers to the spin orientation opposite to σ. Let us introduce the quantities $n = \sum_{\sigma} \int (d\mathbf{p}) n_{\mathbf{p}}(\sigma)$ and $m = \int (d\mathbf{p})(n_{\mathbf{p}}(\uparrow) - n_{\mathbf{p}}(\downarrow))$ corresponding to the total electron density and magnetization respectively. The free energy F takes the form

$$F = \sum_{\sigma} \int (d\mathbf{p})(\epsilon_{\mathbf{p}}(\sigma) - \mu) n_{\mathbf{p}}(\sigma) + \frac{I}{2}(m^2 - n^2), \qquad (2.20)$$

where $\epsilon_{\mathbf{p}}(\uparrow) = \epsilon_{\mathbf{p}}^{(0)} + In - Im$ and $\epsilon_{\mathbf{p}}(\downarrow) = \epsilon_{\mathbf{p}}^{(0)} + In + Im$.

Using the expression for the equilibrium distribution function, Eq. (1.41), together with the definition of the magnetization m, we easily obtain the self-consistent relation

$$m = \int (d\mathbf{p})[\Theta(\mu - \epsilon_{\mathbf{p}}(\uparrow)) - \Theta(\mu - \epsilon_{\mathbf{p}}(\downarrow)]. \qquad (2.21)$$

In addition, the following constraint for n is also satisfied,

$$n = \int (d\mathbf{p})[\Theta(\mu - \epsilon_{\mathbf{p}}(\uparrow)) + \Theta(\mu - \epsilon_{\mathbf{p}}(\downarrow)]. \qquad (2.22)$$

Equations (2.21) and (2.22) are to be selfconsistently solved for the m and the chemical potential μ (the latter is a function of m).

At the onset of ferromagnetism, assuming the magnetization develops continuously from zero, we expand the right hand side of Eq. (2.21) to leading order in m, and ignore the change in μ which appears only at the second order in m. This immediately leads us to the Stoner criterion for ferromagnetic instability:

$$1 = I\nu(0). \qquad (2.23)$$

Chapter 3

Collective Excitations and Zero Sound

So far we have studied various properties of the Fermi liquid in thermodynamic equilibrium. Even in the case of a dynamical phenomenon such as the propagation of a sound wave, we worked in the low frequency limit $\omega\tau \ll 1$; here the quasiparticles had sufficient time to relax before the completion of one period. The condition $\omega\tau \ll 1$ is equivalent to saying that the quasiparticles are able to equilibrate in a region of dimension $l \sim s\tau$ that is much smaller than the wavelength $\lambda = 2\pi s/\omega$ of the sound wave. Such sound waves are a collective, hydrodynamic process. As the temperature is lowered, the relaxation time τ of the quasiparticles increases as $\tau \propto 1/T^2$, and it becomes increasingly difficult to equilibrate the quasiparticles before the completion of a period of oscillation. In other words, at a sufficiently low temperature, or at high enough frequency, the inequality $\omega\tau \ll 1$ can no longer be satisfied, and the hydrodynamic approximation is not appropriate.

When $\omega\tau \gg 1$, sound waves are possible, but they have a different physical character from the hydrodynamic waves. In these waves, collisions of quasiparticles are not important, and neither do we have thermodynamic equilibrium in each volume element. These waves are known as *zero sound*. Zero sound is an essentially nonequilibrium phenomenon. We will study this using a transport equation approach.

The first observation of zero sound was reported in 1963 by Keen and co-workers[1] who were measuring sound attenuation in ^{3}He.

[1] B. E. Keen, P. W. Matthews and J. Wilks, *Physics Letters* **5**, 5 (1963).

3.1 Transport equation for a Fermi liquid

Nonequilibrium states are described by distribution functions that depend not only on the momenta $\mathbf{p}$ but also on the coordinates and time. The nonequilibrium distribution functions $\hat{n}(\mathbf{p}, \mathbf{r}, \boldsymbol{\sigma}, t)$ satisfy the transport equation

$$\frac{d\hat{n}}{dt} = I(\hat{n}), \tag{3.1}$$

where the quantity on the right, $I(\hat{n})$, is known as the *collision integral.* The transport equation describes the change in the number of quasiparticles in a given element of phase space volume due to collisions between them. The change in quasiparticle number can come from two sources. First, it can come from an explicit dependence of $\hat{n}$ on t. Second, it can also come implicitly from the change of the coordinates, momentum, or spin variables $\boldsymbol{\sigma}$ of the quasiparticle in accordance with its equations of motion.

The distinctive feature about the transport equation of a Fermi liquid in comparison with a Fermi gas is that the quasiparticle energy in a Fermi liquid is a functional of the distribution function; i.e., in an inhomogeneous Fermi liquid, besides the distribution $\hat{n}$, the quasiparticle energy also depends on the coordinates.

For small deviations $\delta\hat{n}(\mathbf{p}, \mathbf{r}, \boldsymbol{\sigma}, t)$ from the equilibrium distribution $n_{\mathbf{p}}$, we can write

$$\hat{n}(\mathbf{p}, \mathbf{r}, \boldsymbol{\sigma}, t) = n_{\mathbf{p}} + \delta\hat{n}(\mathbf{p}, \mathbf{r}, \boldsymbol{\sigma}, t). \tag{3.2}$$

Similarly, the departure of the quasiparticle energy $\hat{\epsilon}(\mathbf{p}, \mathbf{r}, \boldsymbol{\sigma}, t)$ from the equilibrium value $\epsilon_{\mathbf{p}}^{(0)}$ can be expressed as

$$\hat{\epsilon}(\mathbf{p}, \mathbf{r}, \boldsymbol{\sigma}, t) = \epsilon_{\mathbf{p}}^{(0)} + \delta\hat{\epsilon}(\mathbf{p}, \mathbf{r}, \boldsymbol{\sigma}, t), \tag{3.3}$$

where

$$\delta\hat{\epsilon}(\mathbf{p}, \mathbf{r}, \boldsymbol{\sigma}, t) = \mathrm{Tr}_{\boldsymbol{\sigma}'} \int (d\mathbf{p}') f(\mathbf{p}, \boldsymbol{\sigma}; \mathbf{p}', \boldsymbol{\sigma}')\delta\hat{n}(\mathbf{p}, \mathbf{r}, \boldsymbol{\sigma}, t). \tag{3.4}$$

The explicitly time-dependent term in the transport equation, Eq.(3.1), is

$$\frac{\partial \hat{n}}{\partial t} = \frac{\partial \delta\hat{n}}{\partial t},$$

while the dependence through the coordinates and momenta gives the contribution

$$\nabla_{\mathbf{r}}\hat{n} \cdot \frac{d\mathbf{r}}{dt} + \nabla_{\mathbf{p}}\hat{n} \cdot \frac{d\mathbf{p}}{dt} = \nabla_{\mathbf{r}}\delta\hat{n} \cdot \frac{d\mathbf{r}}{dt} + \nabla_{\mathbf{p}}\hat{n} \cdot \frac{d\mathbf{p}}{dt}.$$

The Hamilton equations of motion,

$$\frac{d\mathbf{r}}{dt} = \nabla_{\mathbf{p}}\hat{\epsilon}, \qquad \frac{d\mathbf{p}}{dt} = -\nabla_{\mathbf{r}}\hat{\epsilon} = -\nabla_{\mathbf{r}}\delta\hat{\epsilon},$$

allow us to express the above coordinate and momentum dependence of $d\hat{n}/dt$ as

$$\nabla_{\mathbf{r}}\delta\hat{n}\cdot\nabla_{\mathbf{p}}\hat{\epsilon} - \nabla_{\mathbf{p}}\hat{n}\cdot\nabla_{\mathbf{r}}\delta\hat{\epsilon}.$$

Finally, the spin dependence of $d\hat{n}/dt$ is given by the commutator

$$\frac{i}{\hbar}[\hat{\epsilon},\hat{n}].$$

Collecting the above contributions to $d\hat{n}/dt$, we arrive at the following transport equation:

$$\frac{\partial\delta\hat{n}}{\partial t} + \nabla_{\mathbf{r}}\delta\hat{n}\cdot\nabla_{\mathbf{p}}\hat{\epsilon} - \nabla_{\mathbf{p}}\hat{n}\cdot\nabla_{\mathbf{r}}\delta\hat{\epsilon} + \frac{i}{\hbar}[\hat{\epsilon},\hat{n}] = I(\hat{n}). \tag{3.5}$$

We assume for the moment that $\epsilon_{\mathbf{p}}^{(0)}$ and $n_{\mathbf{p}}$ are independent of the spin index. $\delta\hat{n}$ and $\delta\hat{\epsilon}$ are small quantities. Then to the leading order in these small quantities, the transport equation Eq.(3.5) takes the form

$$\frac{\partial\delta\hat{n}}{\partial t} + \nabla_{\mathbf{r}}\delta\hat{n}\cdot\nabla_{\mathbf{p}}\epsilon_{\mathbf{p}}^{(0)} - \nabla_{\mathbf{p}}n_{\mathbf{p}}\cdot\nabla_{\mathbf{r}}\delta\hat{\epsilon} = I(\hat{n}). \tag{3.6}$$

The reason why the commutator makes no contribution to the leading order in $\delta\hat{n}$ and $\delta\hat{\epsilon}$ is that $\epsilon_{\mathbf{p}}^{(0)}$ and $n_{\mathbf{p}}$are independent of the spin index. The commutator is finite only at a higher order, $[\delta\hat{\epsilon},\delta\hat{n}]$. Before solving the transport equation, we need to discuss the conditions for its validity.

3.1.1 Limits of validity of the transport equation

Treating the coordinate and momentum as classical variables (i.e., not quantum operators) amounts to making a quasiclassical (or semiclassical) assumption for the motion of the quasiparticles. The quasiclassical assumption also allows us to define a distribution function that is simultaneously a function of the coordinates and momenta of the quasiparticles. The condition for the validity of the quasiclassical approach is that the de Broglie wavelength of the quasiparticles, $\lambda_F = h/p_F$, is much smaller than the length scale L over which $\hat{n}$ varies appreciably. If $k \sim 1/L$ is the wavevector characterizing the inhomogeneity of $\hat{n}$, then the condition of validity of the transport equation is

$$\hbar k \ll p_F, \text{ or, } \hbar\omega \ll \mu(0) \sim \epsilon_F. \tag{3.7}$$

Since we are interested in understanding the propagation of sound waves in the regime $\omega\tau \gg 1$, we will be working in the following frequency range:

$$\frac{1}{\tau} \ll \omega \ll \frac{\epsilon_F}{\hbar}. \tag{3.8}$$

3.2 Sound propagation at high frequencies

At high enough frequencies, the collision integral $I(\hat{n})$ in the transport equation, Eq.(3.6) is not important. As a first approximation, the collision integral is proportional to $\delta\hat{n}/\tau$, where τ is the relaxation time due to quasiparticle collisions. On the other hand, the time derivative on the left hand side of Eq.(3.6) is proportional to $\omega\delta\hat{n}$, which is much larger than the collision integral on the right hand side. Therefore at high frequencies, we may disregard the collision integral, and the transport equation takes a simpler form

$$\frac{\partial\delta\hat{n}}{\partial t} + \nabla_{\mathbf{r}}\delta\hat{n}\cdot\mathbf{v}_{\mathbf{p}} - \nabla_{\mathbf{p}}n_{\mathbf{p}}\cdot\nabla_{\mathbf{r}}\delta\hat{\epsilon} = 0. \tag{3.9}$$

Here $\mathbf{v}_{\mathbf{p}} = \hat{\mathbf{p}}v_F$, and at $T = 0$, we also have $\nabla_{\mathbf{p}}n_{\mathbf{p}} = -\hat{\mathbf{p}}\delta(p - p_F) = -\hat{\mathbf{p}}v_F\delta(\epsilon_{\mathbf{p}}^{(0)} - \mu(0))$.

Consider first the case of *spin-independent* zero sound. In this case $\delta\hat{n}$ is independent of the index $\boldsymbol{\sigma}$. We seek a solution of the transport equation in the form of a travelling wave,

$$\delta\hat{n}(\mathbf{p},\mathbf{r},t) = \delta(\epsilon_{\mathbf{p}}^{(0)} - \mu(0))\hat{\nu}(\hat{\mathbf{p}})e^{i(\mathbf{k}\cdot\mathbf{r}-\omega t)}. \tag{3.10}$$

Upon substituting the above trial solution in the transport equation Eq.(3.9), we get

$$\begin{aligned}\delta(\epsilon_{\mathbf{p}}^{(0)} - \mu(0))(-\omega + v_F\mathbf{k}\cdot\hat{\mathbf{p}})\hat{\nu}(\hat{\mathbf{p}}) + v_F(\mathbf{k}\cdot\hat{\mathbf{p}})\delta(\epsilon_{\mathbf{p}}^{(0)} - \mu(0))\\ \times\mathrm{Tr}_{\boldsymbol{\sigma}'}\int(d\mathbf{p}')\delta(\epsilon_{\mathbf{p}'}^{(0)} - \mu(0))f\hat{\nu}(\hat{\mathbf{p}}') = 0,\end{aligned} \tag{3.11}$$

which simplifies to

$$(\omega - v_Fk\hat{\mathbf{k}}\cdot\hat{\mathbf{p}})\hat{\nu}(\hat{\mathbf{p}}) = v_Fk(\hat{\mathbf{k}}\cdot\hat{\mathbf{p}})\int\frac{d\Omega'}{4\pi}F^s(\hat{\mathbf{p}},\hat{\mathbf{p}}')\hat{\nu}(\hat{\mathbf{p}}'). \tag{3.12}$$

Let us choose the z direction to lie along the direction $\hat{\mathbf{k}}$, and specify the directions $\hat{\mathbf{p}}$ and $\hat{\mathbf{p}}'$ by the angles (θ,ϕ) and (θ',ϕ') respectively. Let χ be the angle between the directions $\hat{\mathbf{p}}$ and $\hat{\mathbf{p}}'$. We also introduce the dimensionless variable $\lambda = \omega/v_Fk$ and a new function $\tilde{\nu} = (\lambda - \cos\theta)\hat{\nu}$. With this set of notations, the transport equation, Eq.(3.12), takes the final form

$$\tilde{\nu}(\theta,\phi) = \cos\theta\int\frac{d\Omega'}{4\pi}F^s(\chi)\frac{\tilde{\nu}(\theta',\phi')}{\lambda - \cos\theta'}. \tag{3.13}$$

We argue now that in order to have undamped waves, that is, real ω, we must satisfy $\lambda > 1$. Suppose $\lambda < 1$ is a solution. Then the integrand in Eq.(3.13) has a pole at $\lambda = \cos\theta'$. In order to make the integral meaningful, this pole in the plane of the complex variable $\cos\theta'$ needs to be avoided by following a certain rule. That will then impart an imaginary part to the integral, and, therefore, the wave will be damped. Such collisionless damping is known as *Landau damping.* The usual rule for avoiding the pole is to replace λ by $\lambda + i0$. Physically, this shift means that the perturbation remains finite at all previous times, including $t \to -\infty$. $\lambda > 1$ means that the waves of zero sound move faster than the Fermi velocity v_F in the metal. In other words, $\lambda > 1$ is the condition for the quasiparticles to emit Cerenkov waves of zero sound.

Eq.(3.13) can be analytically solved for $\tilde{\nu}$ and λ in some simple cases. If $F^s(\chi)$ is a constant F_0^s, the integral on the right hand side of Eq.(3.13) is independent of θ, ϕ; thus, the solution $\tilde{\nu}$ has the form

$$\tilde{\nu} = \hat{\nu}(\lambda - \cos\theta) = \text{const.} \times \cos\theta. \tag{3.14}$$

This means for *zero sound* waves, the Fermi surface is a surface of revolution elongated in the forward direction of the wave's propagation and flattened in the opposite direction. This anisotropy is a consequence of the non-equilibrium state of the liquid. In the equilibrium state, the Fermi surface is a sphere. To compare with the ordinary sound waves, the Fermi surface in that case remains spherical with an oscillating radius (p_F varies with the density), and the sphere is shifted by an amount that depends on the velocity of the liquid in the wave. Thus for *ordinary sound* waves, $\hat{\nu} = \delta p_F + \text{const.} \times \cos\theta$.

To complete the solution, we need the value of λ. Using the solution of Eq.(3.14) in Eq.(3.13), we have

$$1 = \frac{F_0^s}{2} \int_0^\pi d\theta \frac{\sin\theta \cos\theta}{\lambda - \cos\theta}. \tag{3.15}$$

Performing the integration we find that λ is a solution of the following equation:

$$\frac{1}{F_0^s} = \frac{\lambda}{2} \ln\left[\frac{\lambda + 1}{\lambda - 1}\right] - 1. \tag{3.16}$$

The right hand side of Eq.(3.16) is always positive for $\lambda > 1$. A real solution for λ exists only if $F_0^s > 0$. When the interaction parameter $F_0^s \to 0$, the solution for λ is

$$\lambda = \frac{\omega}{v_F k} \approx 1 + \frac{2}{e^2} e^{-2/F_0^s}. \tag{3.17}$$

Thus in a Fermi liquid with weak interparticle interaction, the speed of zero sound $\approx p_F/m^* \approx p_F/m$ is larger than the speed of ordinary sound $s = p_F(1 + F_0^s)/\sqrt{3mm^*} \approx p_F/\sqrt{3}m$ by a factor of $\sqrt{3}$. At large values of the interaction parameter, the solution for λ is

$$\lambda \approx \sqrt{F_0^s/3}. \tag{3.18}$$

3.3 Zero sound involving spin

Suppose $\delta\hat{n}$ does depend on its spin index. In that case let us try a spin-dependent zero sound solution for the transport equation, Eq.(3.11):

$$\hat{\nu} = \boldsymbol{\sigma} \cdot \hat{\mu}. \tag{3.19}$$

Following the same procedure as for the spin-independent case, and introducing the new variable $\tilde{\mu} = (\lambda - \cos\theta)\hat{\mu}$, we get the final form of the transport equation:

$$\tilde{\mu}(\theta, \phi) = \cos\theta \int \frac{d\Omega'}{4\pi} F^a(\chi) \frac{\tilde{\mu}(\theta', \phi')}{\lambda - \cos\theta'}. \tag{3.20}$$

This leads to similar solutions for the speed of the spin-dependent zero sound as in Eq.(3.16) except that now F_0^s should be replaced by F_0^a. Again, undamped solutions (i.e. real λ) are possible only if $F_0^a > 0$. In liquid ^{3}He, it is observed that $F_0^a < 0$; therefore, undamped spin-dependent waves of zero sound cannot be propagated. On the other hand, $F_0^s > 0$ in liquid ^{3}He making it possible to have undamped waves of zero sound.

3.4 Damped zero sound

We have seen that undamped zero sound is possible if $F_0^{s(a)} > 0$. Let us study what happens when, say, $F^s < 0$. Then Eq.(3.16),

$$-\frac{1}{|F_0^s|} = \frac{\lambda}{2} \ln\left[\frac{\lambda+1}{\lambda-1}\right] - 1,$$

cannot give us purely real solutions for λ. If $F_0^s \ll -1$, then the solution for λ is

$$\lambda \approx \pm i\sqrt{|F_0^s|/3}, \tag{3.21}$$

that is, the zero sound is unstable. The system will build up permanent density fluctuations that will eventually be limited by nonlinear effects. We can also show, more generally, that the solutions for λ are purely imaginary if $F_0^s < -1$. Note that this condition for instability is identical to what we obtained in Chapter 2 on Pomeranchuk stability conditions. For $-1 < F_0^s < 0$, λ is complex. That means waves of zero sound exist, but they are (Landau) damped.

Chapter 4

Scattering Theory

We will now take a brief break from Fermi liquids and devote some time to recapitulate a few topics in scattering theory. We need this in order to understand the physics discussed in the next Chapter which concerns the Fermi liquid properties of a system consisting of a single magnetic impurity immersed in a sea of conduction electrons. Scattering of the conduction electrons from the magnetic impurity generates an extra interaction between the electrons which can be understood using Fermi liquid theory.

A scattering process is said to be elastic if the internal properties of the scattering entities remain the same upon collision and inelastic if the internal properties change upon collision. By analyzing the scattering properties, we can get information about the interaction of the scattering entities.

4.1 Scattering phase shift

We now study a simple model of a particle scattering from a spherically symmetric square well potential. This will acquaint us with the concept of a scattering phase shift. Consider the Schrödinger equation,

$$\left[-\frac{\hbar^2}{2m}\nabla^2 + V(r)\right]\psi = E\psi, \tag{4.1}$$

of a particle moving in a spherically symmetric square well potential $V(r)$ given by

$$V(r) = \begin{cases} -V_0, & r \leq a, \\ 0, & r > a. \end{cases} \tag{4.2}$$

For a spherically symmetrical potential, the wave function ψ will be of the form

$$\psi(\mathbf{r}) = R_l(r)Y_{lm}(\theta, \phi), \tag{4.3}$$

where the Y_{lm} are the spherical harmonics and the $R_l(r)$ satisfy

$$\left[-\frac{\hbar^2}{2m}\left(\frac{d^2}{dr^2} + \frac{2}{r}\frac{d}{dr} - \frac{l(l+1)}{r^2}\right) + V(r)\right] R_l(r) = ER_l(r). \tag{4.4}$$

Define the wavevectors $k' = \sqrt{2m(E+V_0)}/\hbar$ and $k = \sqrt{2mE}/\hbar$. For $r \le a$, using the boundary condition at $r = 0$ gives

$$R_l(r) = A_l j_l(k'r), \tag{4.5}$$

and for $r > a$, the general solution of Eq.(4.4) is

$$R_l(r) = B_l j_l(kr) + C_l n_l(kr), \tag{4.6}$$

where j_l is the spherical Bessel function of order l and n_l is the spherical Neumann function of order l. The first few have fairly simple forms:

$$\begin{aligned} j_0(x) &= \sin x/x, \qquad & j_1(x) &= \sin x/x^2 - \cos x/x, \\ n_0(x) &= -\cos x/x, \qquad & n_1(x) &= -\cos x/x^2 - \sin x/x. \end{aligned} \tag{4.7}$$

The asymptotic forms for j_l and n_l for $kr \gg l$ are

$$\begin{aligned} j_l(kr) &\sim \frac{1}{kr}\sin(kr - l\pi/2), \\ n_l(kr) &\sim -\frac{1}{kr}\cos(kr - l\pi/2), \end{aligned} \tag{4.8}$$

while close to the origin, j_l has the form

$$j_l(kr) = \frac{(kr)^l}{(2l+1)!!}\left[1 - \frac{(kr)^2}{2(1-2l)} + \cdots\right]. \tag{4.9}$$

Introducing the ratio

$$\frac{C_l}{B_l} = -\tan\delta_l(k), \tag{4.10}$$

the radial wavefunction, Eq.(4.6), at large kr may be re-expressed as

$$R_l(r)\big|_{r\to\infty} = \frac{B_l}{\cos\delta_l(k)}\frac{1}{kr}\sin(kr - l\pi/2 + \delta_l(k)). \tag{4.11}$$

$\delta_l(k)$ is known as the *phase shift*. In the absence of a scattering potential $V(r)$, the phase shift would have been zero. $\delta_l(k)$ contains information about

$V(r)$. Sometimes, instead of writing the $r > a$ solution in terms of j_l and n_l, the spherical Hankel functions $h_l^{(1)} = j_l + in_l$ and $h_l^{(2)} = [h_l^{(1)}]^*$ are used. The advantage is that at large distances $h_l^{(1)}(kr) \sim (1/ikr)e^{i(kr-l\pi/2)}$ and $h_l^{(2)}(kr)$, respectively, look like spherical waves going away from the origin and coming towards the origin. In this notation, the asymptotic solution, Eq.(4.11), is proportional to $h_l^{(2)}(kr) + e^{i2\delta_l(k)}h^{(1)}(kr)$. Thus $\delta_l(k)$ is the extra phase the outgoing particle picks up upon scattering.

The boundary conditions for the wavefunction are that R_l as well as its first derivative are continuous everywhere. Therefore we may write

$$\left.\frac{(dj_l(k'r)/dr)}{j_l(k'r)}\right|_{r=a} = \left.\frac{(dj_l(kr)/dr) + C_l/B_l(dn_l(kr)/dr)}{j_l(kr) + C_l/B_l n_l(kr)}\right|_{r=a}. \tag{4.12}$$

Eq.(4.12) gives us the ratio $C_l/B_l = -\tan\delta_l(k)$. For $s-$wave scattering (i.e. $l = 0$), the wavefunction, given by Eq.(4.7), is particularly simple, and using the above boundary condition we get

$$k' \cot(k'a) = k \cot(ka + \delta_0(k)). \tag{4.13}$$

We now introduce the concept of the scattering cross section, σ, and the scattering function, $f(\theta, \phi)$.

4.2 Cross section and scattering function

One way to measure scattering phenomena is to direct a beam of particles at the target and count the number of particles scattered into each elemental solid angle $d\Omega$ subtended at the target. This can be made formal by introducing a differential scattering cross section $d\sigma/d\Omega$:

$$\begin{aligned}\frac{d\sigma}{d\Omega}d\Omega &= \frac{\text{number of particles scattered into } d\Omega \text{ per unit time}}{\text{number of incident particles per unit area per unit time}} \\ &= \frac{r^2 d\Omega \times \text{scattered flux}}{\text{incident flux}}.\end{aligned} \tag{4.14}$$

The total cross section is $\sigma_{\text{tot}} = \int d\Omega(d\sigma/d\Omega)$.

We can assume the z direction of the coordinate system to be along the direction of the incident flux which can be regarded as an incoming plane wave, $\psi_{in} = e^{ikz}$, from the direction $z = -\infty$. The scattering potential is centered at the origin. After the scattering process, we get, in addition to the outgoing plane wave, a spherical wave emanating from the scattering region. We thus try a solution of the Schrödinger equation which in the asymptotic limit (i.e., far from the scattering region) has the form

$$\psi = e^{ikz} + f(\theta, \phi)\frac{e^{ikr}}{r}. \tag{4.15}$$

$f(\theta, \phi)$ is known as the scattering function.[1] It is related to the differential scattering cross section through $|f(\theta, \phi)|^2 = d\sigma/d\Omega$.

4.3 Partial wave expansion

We now develop a formal way to describe scattering from an arbitrary short ranged spherically symmetric potential $V(r)$ at the origin using the ideas of the scattering phase shift, cross section and the scattering function we have introduced in the previous sections. By "short ranged" we mean

$$\left| \int_0^\infty dr\, r^2 V(r) \right|^2 < \infty. \tag{4.16}$$

The physical reason for such a definition of "short ranged" will become clear as we further develop our theory.

The Schrödinger equation for a particle moving in such a potential $V(r) = \hbar^2 U(r)/2m$ can be written in the form

$$[\nabla^2 + k^2 - U(r)]\psi = 0, \tag{4.17}$$

where $k = \sqrt{2mE}/\hbar$ as before. We decompose $\psi(\mathbf{r})$ into spherical partial waves,[2]

$$\psi(\mathbf{r}) = \sum_{l=0}^{\infty} i^l (2l+1) R_l(r) P_l(\cos\theta). \tag{4.18}$$

$R_l(r)$ satisfies the differential equation

$$\left(\frac{d^2}{dr^2} + \frac{2}{r}\frac{d}{dr} - \frac{l(l+1)}{r^2} + k^2 - U(r) \right) R_l(r) = 0. \tag{4.19}$$

Because of the condition in Eq.(4.16), $U(r)$ cannot decrease faster than $1/r^2$ near the origin. Hence, for r near the origin, we can ignore the potential term and solve the differential equation for R_l. We have already seen that the solution is proportional to j_l which evolves as r^l near the origin. Thus $R_l \propto r^l$ can be regarded as a boundary condition near $r = 0$. The other

---

[1] We do not intend to suggest here a relation between the Fermi liquid parameter f and the scattering function.

[2] In general we should have resolved in spherical harmonics $Y_{lm}(\theta, \phi)$; however, the incident wave has no dependence on ϕ, and since we are studying elastic scattering, the scattered wave should also show no dependence on ϕ. For if the scattered wave did have a $\phi-$dependence, that would imply a change in the magnetic quantum number of the scattering particle which is not consistent with the definition of elastic scattering.

boundary condition will depend on the particular details of the potential. We have already shown that for a square well potential, the radial part of the wavefunction for large values of r, has the characteristic form $R_l \propto (h_l^{(2)}(kr) + e^{i2\delta_l(k)} h_l^{(1)}(kr))$, and the $\delta_l(k)$ are determined by the boundary conditions.

Resolving the incident plane wave in a partial wave series similar to Eq.(4.18) will enable us to make a term by term comparison between the incident and scattered wave:

$$\begin{aligned} e^{ikz} = e^{ikr\cos\theta} &= \sum_{l=0}^{\infty} i^l (2l+1) j_l(kr) P_l(\cos\theta) \\ &= \frac{1}{2}\sum_{l=0}^{\infty} i^l (2l+1)(h_l^{(1)}(kr) + h_l^{(2)}(kr)) P_l(\cos\theta). \end{aligned} \tag{4.20}$$

Remember we had identified $h_l^{(1)}$ as a spherical wave going away from the origin and $h_l^{(2)}$ as a spherical wave coming towards the origin. Not surprisingly, the plane wave has equal amplitudes for the incoming and outgoing components.

In the presence of a short range scatterer at the origin, the outgoing wave will get modified by a factor $S_l(k)$. So we may express the wavefunction, Eq.(4.18), as

$$\psi(\mathbf{r})\Big|_{r\to\infty} = \frac{1}{2}\sum_{l=0}^{\infty} i^l (2l+1)[h_l^{(2)}(kr) + S_l(k) h_l^{(1)}(kr)] P_l(\cos\theta). \tag{4.21}$$

For the square well potential, we had seen that $S_l(k)$ was simply a phase factor, $e^{i2\delta_l(k)}$. This is true for any *elastic* scattering process. If the scattering is elastic, then the flux of the outgoing wave cannot differ from that of the incoming wave. Further, because of the conservation of angular momentum, this should be the case for each partial wave.

Using Eq.(4.20) for the resolution of the incident plane wave, we can separate out the incident wave in Eq.(4.21) and re-express the solution as

$$\psi(\mathbf{r})\Big|_{r\to\infty} = e^{ikz} + \frac{1}{2}\sum_{l=0}^{\infty} i^l (2l+1)(e^{i2\delta_l(k)} - 1) h_l^{(1)}(kr) P_l(\cos\theta). \tag{4.22}$$

Using the asymptotic form $h_l^{(1)}(kr) \approx e^{i(kr - l\pi/2)}/ikr = i^{-l} e^{ikr}/ikr$, and the ansatz Eq.(4.15), we can immediately read off the expression for the scattering function $f(\theta)$:

$$f(\theta) = \sum_{l=0}^{\infty} f_l P_l(\cos\theta) = \frac{1}{k}\sum_{l=0}^{\infty}(2l+1) e^{i\delta_l(k)} \sin\delta_l(k)\, P_l(\cos\theta). \tag{4.23}$$

Here we have introduced f_l which is the lth angular momentum component of $f(\theta)$. The phase shift $\delta_l(k)$ has a simple relation with f_l :

$$\delta_l(k) = \arg(f_l). \tag{4.24}$$

Often we have a direct means of obtaining the f_l. Eq.(4.24) can then be used to get the phase shifts.

Since $d\sigma/d\Omega = |f(\theta)|^2$, we can easily obtain the total cross section,

$$\sigma_{\text{tot}} = \sum_{l=0}^{\infty} \sigma_l = \frac{4\pi}{k^2} \sum_{l=0}^{\infty} (2l+1) \sin^2 \delta_l(k). \tag{4.25}$$

For this, we would need to use the orthogonality relation for Legendre polynomials,

$$\int_0^{\pi} d\Omega P_l(\cos\theta) P_{l'}(\cos\theta) = \frac{4\pi}{(2l+1)} \delta_{ll'}. \tag{4.26}$$

For elastic scattering, the maximum possible value of the σ_l is $(4\pi/k^2)(2l+1)$. That happens whenever

$$\delta_l = (N + 1/2)\pi \tag{4.27}$$

is satisfied. The upper limit for σ_l is called the *unitarity bound.* Whenever the resonance condition, Eq.(4.27), is satisfied, the scattering cross section in that (angular momentum) channel shows a maximum.

So far we have solved only one simple case (in Eq.(4.13)) where we explicitly related the phase shift $\delta_0(k)$ to the scattering potential. In the next section we formulate a general approach for calculating $f(\theta)$. By resolving $f(\theta)$ into a series in $P_l(\cos\theta)$, we can work out the various phase shifts through Eq.(4.23).

4.4 Born approximation

Let us write the Hamiltonian as the sum of a "free" part $\hat{H}_0$ and an elastic scattering potential $\hat{V}$,

$$\hat{H} = \hat{H}_0 + \hat{V} = \frac{\hat{\mathbf{p}}^2}{2m} + \hat{V}. \tag{4.28}$$

Let $|\psi^{(0)}\rangle$ be an eigenstate of $\hat{H}_0$ corresponding to energy eigenvalue E. Since the scattering is elastic, E will also be an eigenvalue of $\hat{H}$. Let $|\psi\rangle$ be the eigenstate of $\hat{H}$ corresponding to the eigenvalue E. The solution for $(\hat{H}_0 + \hat{V})|\psi\rangle = E|\psi\rangle$ can be formally written as $|\psi\rangle = |\chi\rangle + (E - \hat{H}_0 + i\eta)^{-1}\hat{V}|\psi\rangle$, where $|\chi\rangle$ satisfies $(E - \hat{H}_0)|\chi\rangle = 0$, and η is an infinitesimally small positive number. As $\hat{V} \to 0$, we would expect $|\psi\rangle$ to converge to $|\psi^{(0)}\rangle$. Therefore, the formal solution is

$$|\psi\rangle = |\psi^{(0)}\rangle + \frac{1}{(E - \hat{H}_0 + i\eta)} \hat{V}|\psi\rangle. \tag{4.29}$$

Eq.(4.29) is called the Lippmann-Schwinger equation. The wavefunctions of the Schrödinger equation are position representations of the eigenstates. For example, $\langle \mathbf{r}|\psi^{(0)}\rangle = \psi^{(0)}(\mathbf{r}) = e^{i\mathbf{k}\cdot\mathbf{r}}$ is the free particle wavefunction corresponding to energy E. In the position representation, Eq.(4.29) becomes

$$\begin{aligned}\langle \mathbf{r}|\psi\rangle &= \langle \mathbf{r}|\psi^{(0)}\rangle + \langle \mathbf{r}|\frac{1}{(E-\hat{H}_0+i\eta)}\hat{V}|\psi\rangle \\ &= \langle \mathbf{r}|\psi^{(0)}\rangle + \int d\mathbf{r}'\langle \mathbf{r}|\frac{1}{(E-\hat{H}_0+i\eta)}|\mathbf{r}'\rangle\langle \mathbf{r}'|\hat{V}|\psi\rangle. \end{aligned} \tag{4.30}$$

The quantity $\langle \mathbf{r}|\frac{1}{(E-\hat{H}_0+i\eta)}|\mathbf{r}'\rangle$ is related to the Green function $G^{(0)}(\mathbf{r},\mathbf{r}')$of the Helmholtz equation, $(\nabla^2+k^2)G^{(0)}(\mathbf{r},\mathbf{r}') = \delta(\mathbf{r}-\mathbf{r}')$:

$$G^{(0)}(\mathbf{r},\mathbf{r}') = \frac{\hbar^2}{2m}\langle \mathbf{r}|\frac{1}{(E-\hat{H}_0+i\eta)}|\mathbf{r}'\rangle = -\frac{1}{4\pi}e^{ik|\mathbf{r}-\mathbf{r}'|}. \tag{4.31}$$

Here $k = \sqrt{2mE}/\hbar$.

Suppose V is centered at $\mathbf{r}' = 0$. Since we want to know the wavefunction at a point $\mathbf{r}$ that is far from the scattering region, we can approximate

$$|\mathbf{r}-\mathbf{r}'| \simeq (r^2 - 2\mathbf{r}\cdot\mathbf{r}')^{1/2} \simeq r - \hat{\mathbf{r}}\cdot\mathbf{r}'. \tag{4.32}$$

Using Eq.(4.31) and Eq.(4.32) in Eq.(4.30) we get

$$\psi(\mathbf{r}) = e^{i\mathbf{k}\cdot\mathbf{r}} - \frac{m}{2\pi\hbar^2}\frac{e^{ikr}}{r}\int d\mathbf{r}' e^{-ik\hat{\mathbf{r}}\cdot\mathbf{r}'}V(\mathbf{r}')\psi(\mathbf{r}'). \tag{4.33}$$

Let $\mathbf{k}' = k\hat{\mathbf{r}}$. In Eq.(4.33) we can now read off the scattering function,

$$f(\mathbf{k}',\mathbf{k}) = -\frac{m}{2\pi\hbar^2}\int d\mathbf{r}' e^{-i\mathbf{k}'\cdot\mathbf{r}'}V(\mathbf{r}')\psi(\mathbf{r}'). \tag{4.34}$$

In order to solve Eq.(4.34) for f, we make a perturbation expansion of $\psi(\mathbf{r}')$ in powers of V. The first term in the expansion is called the first Born approximation, and so on. The set of such terms in ascending powers of V is known as the Born series. To leading order in V, we can say $\psi(\mathbf{r}') \approx \psi^{(0)}(\mathbf{r}') = e^{i\mathbf{k}\cdot\mathbf{r}'}$. Thus the first Born approximation for f is

$$f^{(1)}(\mathbf{k}',\mathbf{k}) = -\frac{m}{2\pi\hbar^2}\int d\mathbf{r}' e^{i(\mathbf{k}-\mathbf{k}')\cdot\mathbf{r}'}V(\mathbf{r}') = -\frac{m}{2\pi\hbar^2}\langle \mathbf{k}'|\hat{V}|\mathbf{k}\rangle. \tag{4.35}$$

4.4.1 The T-matrix

The Born series can be neatly expressed in terms of the $T-$matrix which is defined as

$$\hat{T}|\psi^{(0)}\rangle = \hat{V}|\psi\rangle. \tag{4.36}$$

This allows us to write the Lippmann-Schwinger equation in terms of only the $|\psi^{(0)}\rangle$ states:

$$\hat{T}|\psi^{(0)}\rangle = \hat{V}|\psi^{(0)}\rangle + \hat{V}\frac{1}{(E - \hat{H}_0 + i\eta)}\hat{T}|\psi^{(0)}\rangle, \tag{4.37}$$

or,

$$\hat{T} = \hat{V} + \hat{V}\frac{1}{(E - \hat{H}_0 + i\eta)}\hat{T}. \tag{4.38}$$

The solution of Eq.(4.38) is

$$\hat{T} = \hat{V} + \hat{V}\frac{1}{(E - \hat{H}_0 + i\eta)}\hat{V} + \hat{V}\frac{1}{(E - \hat{H}_0 + i\eta)}\hat{V}\frac{1}{(E - \hat{H}_0 + i\eta)}\hat{V} + \cdots. \tag{4.39}$$

The scattering function has a simple relation with the $T-$matrix:

$$f(\mathbf{k}', \mathbf{k}) = -\frac{m}{2\pi\hbar^2}\langle\mathbf{k}'|\hat{T}|\mathbf{k}\rangle. \tag{4.40}$$

A leading order (in V) approximation to $\hat{T}$ in Eq.(4.40) will give us the first term in the Born series (i.e., Eq.(4.35)).

4.5 Friedel sum rule

Consider a large spherical cavity of size R, and assume the cavity wall is hard. The radial part of the wave function of the particle is, as for the free particle, $R_l(r) = j_l(kr)$. However now the finiteness of the system permits only a discrete set of k. Only those values of k are permitted that satisfy $j_l(kR) = 0$. From the known asymptotic form of j_l we showed in Eq.(4.8), we gather that the allowed values of k are approximately given by

$$k_{ln}^{(0)} \simeq \frac{1}{R}\left(\frac{l\pi}{2} + n\pi\right), \qquad k_{ln}^{(0)}R \gg \frac{l\pi}{2}, \tag{4.41}$$

where n is a natural number. Thus for a given l, we add a new mode each time $n \to n+1$, i.e., the quantity in brackets in the above equation changes by π. If we now introduce a scattering potential at the center of the sphere, then we can similarly show that the allowed values of k are given by

$$k_{ln} \simeq \frac{1}{R}\left(\frac{l\pi}{2} + n\pi - \delta_l(k_{ln})\right), \qquad k_{ln}R \gg \frac{l\pi}{2}. \tag{4.42}$$

Therefore, for a fixed l, n, the change in the number of modes compared to the case where there was no scatterer is $-\delta_l(k_{ln})/\pi$. So if $\delta_l(k_{ln}) = \pi$, we can see that there will be one extra mode up to k_{ln}. Physically, this extra mode is a bound state of the (attractive) scattering potential. The change in the total number of modes between two fixed values k_1 and k_2 is found by summing over the contribution from all l and spins σ :

$$\frac{1}{\pi}\sum_{\sigma}\sum_{l=0}^{\infty}(2l+1)(\delta_l(k_2) - \delta_l(k_1)). \tag{4.43}$$

The total extra charge introduced into the system because of the presence of the scattering potential is

$$\Delta N = \frac{1}{\pi}\sum_{\sigma}\sum_{l=0}^{\infty}(2l+1)\delta_l(k). \tag{4.44}$$

Eq.(4.44) is known as the *Friedel sum rule.* This sum rule will prove very useful to us in the next chapter. An alternate version of the Friedel sum rule can be written by using Eq.(4.24) and Eq.(4.40) in Eq.(4.44):

$$\Delta N = \frac{1}{\pi}\mathrm{Im}\ln[\det(T(k))]. \tag{4.45}$$

Chapter 5

Fermi Liquid Regime of the Kondo Problem

A localized particle with spin S in thermal equilibrium with a nonmagnetic bath and subjected to a magnetic field will have an average polarization,

$$M_z(B_{\text{ext}}, T) = (g\mu_B)\frac{\sum_{m=-S}^{S} m \exp(mg\mu_B B_{\text{ext}}/T)}{\sum_{m=-S}^{S} \exp(mg\mu_B B_{\text{ext}}/T)}. \tag{5.1}$$

For small fields such that $g\mu_B B_{\text{ext}}/T \ll 1$, the polarization obeys the Curie law:

$$M_z = \chi_{\text{imp}}(T) B_{\text{ext}} \equiv (g\mu_B)^2 \frac{S(S+1)}{3T} B_{\text{ext}}. \tag{5.2}$$

On the other hand, if we consider a degenerate (i.e. $T \ll \epsilon_F$) free electron gas at a density n, the polarization in small magnetic fields is given by the Pauli spin susceptibility:

$$M_z = \chi B_{\text{ext}} \equiv \frac{g}{2} g\mu_B^2 \nu(0) B_{\text{ext}}, \tag{5.3}$$

where $\nu(0) = 3n/2\epsilon_F$ is as usual the density of states of electrons at the Fermi energy, and n is the electron density.

If a magnetic impurity is introduced into a conduction electron bath, will it continue to follow a Curie law for the impurity susceptibility? Or will the properties of the impurity be more like the host which has a Pauli susceptibility? This is the basic issue in Kondo physics.

When the magnetic impurity is introduced into the metal, the susceptibility is not just the sum of χ_{imp} and χ, because, the impurity and the conduction electrons in the metal will in general mutually interact. A single impurity cannot overwhelm the behavior of the large number of electrons in the metal. So the natural question we should pose is the extent to which the impurity susceptibility deviates from the Curie law. Does it, at any stage, get subsumed into the sea of conduction electrons resulting in a constant, Pauli-like susceptibility at the lowest temperatures? These questions collectively describe the Kondo problem.

In this chapter, we are mostly interested in the behavior at very low temperatures. Nevertheless, it is appropriate to first present a brief overview of the physical problem, including behavior at not very low temperatures.

5.1 The Kondo effect

Effects of magnetic impurities on the properties of metals have been known since the 1930s. In chemically pure metal samples (i.e. with insignificant impurity concentration), one generally observes that the resistance decreases as the temperature is lowered because, at lower temperatures, there are fewer phonons to scatter the conduction electrons. This is also the observation in somewhat impure samples at not very low temperatures where scattering by phonons is still dominant over other mechanisms. When some impure samples are cooled further, as for example gold with iron impurities, it was observed that at a low enough temperature, the resistance goes through a minimum and then begins to rise.

In 1959, Alekseevsky and Gadyukov found that the resistance upturn has the dependence $\ln(1/T)$. The explanation for this temperature dependence was provided by Kondo in 1964 and the effect was named after him. We briefly discuss Kondo's approach.

Kondo considered a localized spin$-1/2$ magnetic impurity interacting with the conduction electrons through an *antiferromagnetic* exchange interaction. The Hamiltonian of such a system can be written as

$$H = \sum_{\mathbf{k},\alpha} \epsilon_{\mathbf{k}} c^{\dagger}_{\mathbf{k}\alpha} c_{\mathbf{k}\alpha} + J \sum_{\mathbf{k},\mathbf{k}'} \sum_{i,\alpha\beta} c^{\dagger}_{\mathbf{k}\alpha} \sigma^{i}_{\alpha\beta} c_{\mathbf{k}'\beta} \cdot S^{i}, \tag{5.4}$$

where the first term refers to the band of conduction electrons, and in the second term, S^i are the spin operators corresponding to the components of the impurity spin $\mathbf{S}$. σ^i are the Pauli spin matrices. $J > 0$ is the strength of the antiferromagnetic coupling between the magnetic impurity and the conduction electrons at the site of the impurity.

The magnetic impurity scatters conduction electrons impinging on it which affects the resistance of the sample. The contribution to the resistivity, ρ_{imp}, is proportional to the impurity scattering rate $1/\tau_{\text{imp}}$. For calculating $1/\tau_{\text{imp}}$, we can use the techniques developed in Chapter 4. First we estimate the scattering function in terms of the $T-$matrix. The scattering rate at any wavevector (and angular momentum) is directly proportional to the corresponding differential scattering cross section. The relation between the scattering rate and the

$T-$matrix is

$$\frac{1}{\tau(k)} = \frac{2\pi}{\hbar} c_{\text{imp}} \int (d\mathbf{k}')\delta(\epsilon_{\mathbf{k}} - \epsilon_{\mathbf{k}'})|T_{\mathbf{k}\mathbf{k}'}|^2(1 - \cos\chi). \tag{5.5}$$

Here c_{imp} is the impurity concentration and χ is the angle between the directions of $\mathbf{k}$ and $\mathbf{k}'$.

The particular temperature dependence of the scattering, $\ln(1/T)$, can be obtained by making a perturbation expansion for the $T-$matrix in powers of the small parameter $J\nu(0)$ [see Eq.(4.39)]. A detailed microscopic calculation will be presented later in this book.We just present the relevant results here.

The leading term in the $T-$matrix is

$$T^{(1)}_{\mathbf{k}\alpha\sigma;\mathbf{k}'\alpha\sigma'} = J \sum_{i=1}^{3} \sigma^i_{\alpha\alpha'} S^i_{\sigma\sigma'} \tag{5.6}$$

and the contribution of this term to the impurity resistivity is

$$\rho^{(1)}_{\text{imp}} = \frac{3\pi m J^2 \nu(0) c_{\text{imp}}}{4ne^2\hbar}. \tag{5.7}$$

Clearly, there is no temperature dependence here. Kondo went on to estimate the second order (in J) contribution to the $T-$matrix:

$$T^{(2)}_{\mathbf{k}\alpha\sigma;\mathbf{k}'\alpha\sigma'} = J^2\nu(0)\ln(D/k_BT) \sum_{i=1}^{3} \sigma^i_{\alpha\alpha'} S^i_{\sigma\sigma'}, \tag{5.8}$$

where D is the (band) width of the conduction electron band. It is of the order of the Fermi energy. The contribution of this term to the impurity resistivity is

$$\rho^{(2)}_{\text{imp}} = \rho^{(1)}_{\text{imp}} \times 2J\nu(0)\ln(D/k_BT). \tag{5.9}$$

The second order contribution does reveal a logarithmic temperature dependence. Kondo's perseverance was worth it. Similarly we can obtain the temperature dependence of the impurity susceptibility and specific heat. The final results are given below:

$$\begin{aligned}
\rho_{\text{imp}} &\approx \frac{3\pi m J^2 \nu(0) c_{\text{imp}}}{4ne^2\hbar}\left[1 + 2J\nu(0)\ln\left(\frac{D}{k_BT}\right)\right], \\
\chi_{\text{imp}} &\approx \frac{(g_s\mu_B)^2 c_{\text{imp}}}{4k_BT}\left[1 - J\nu(0) - (J\nu(0))^2\ln\left(\frac{D}{k_BT}\right)\right], \\
C_{\text{imp}} &\approx \frac{3\pi^2 k_B (J\nu(0))^4 c_{\text{imp}}}{4}\left[1 + 4J\nu(0)\ln\left(\frac{D}{k_BT}\right)\right].
\end{aligned} \tag{5.10}$$

All these logarithmic corrections to various properties may be captured succinctly by introducing the idea of an effective coupling, J_{eff} :

$$J_{\text{eff}}\nu(0) = J\nu(0) + (J\nu(0))^2\ln(D/k_BT) + O(J^3), \tag{5.11}$$

thus, we can say

$$\rho_{\text{imp}} = \frac{3\pi m\nu(0)c_{\text{imp}}}{4ne^2}J_{\text{eff}}^2, \tag{5.12}$$

and

$$\chi_{\text{imp}} = \frac{(g_s\mu_B)^2 c_{\text{imp}}}{4k_BT}(1 - J_{\text{eff}}\nu(0)). \tag{5.13}$$

In fact, one can sum to all orders in J, keeping the most logarithmically divergent terms in $\ln(D/k_BT)$ at each order in the perturbation theory, and the result of the summation can be shown to be

$$J_{\text{eff}} = \frac{J}{1 - J\nu(0)\ln(D/k_BT)}. \tag{5.14}$$

Thus J_{eff} increases continuously as the temperature is decreased, and diverges at a low temperature T_K, where

$$T_K = De^{-1/J\nu(0)}. \tag{5.15}$$

T_K is called the Kondo temperature. One can rewrite the effective coupling in Eq.(5.14) as

$$J_{\text{eff}}\nu(0) = \frac{1}{\ln(T/T_K)}. \tag{5.16}$$

One naturally wonders what is the physical meaning of the divergence of the effective coupling at the Kondo temperature, and what are the physical consequences of this divergence. Does the impurity contribution to the resistivity increase without bounds as the temperature is lowered, or, does it ultimately saturate at a finite value? Similarly, are we to expect, as Eq.(5.13) would suggest, that the impurity susceptibility becomes small and ultimately disappears at the Kondo temperature? What is the behavior below the Kondo temperature? These questions cannot be answered using perturbation theory because what we are witnessing right now is a breakdown of our perturbative approach. Equations 5.12 and 5.13 are not valid in the vicinity of the Kondo temperature. Understanding the physics below the Kondo temperature requires a different approach and that will be the subject of the remaining part of this chapter.

5.2 Physics below the Kondo temperature

5.2.1 Failure of perturbative approaches

We have seen that the antiferromagnetic interaction $J_{\rm eff}$ between the impurity and the conduction electrons grows and becomes infinitely large as we approach the Kondo temperature from above. An infinitely strong antiferromagnetic interaction means that the $S = 1/2$ impurity will form a tightly-bound singlet with a conduction electron. Once the magnetic character of the impurity is thus lost, we end up with a simple potential scatterer. This would suggest, for instance, that the impurity resistivity should ultimately saturate at $T = 0$.

One shortcoming of the summation of the perturbation series we described in the last section is that we only selected the most strongly divergent terms at each order in J. So the first strategy we could take in dealing with our problems near $T = T_K$ is to retain in our perturbation series other terms that are not as strongly divergent. In general it is impossible to perform this sum because the number of diagrams rises very rapidly with the order of expansion in J. Subsequently, some indirect methods of effectively performing such sums were formulated by Suhl, Abrikosov and others around 1965. The following expression for the impurity resistivity was provided by Hamann in 1967:

$$\rho_{\rm imp} = \frac{\rho_0}{2}\left[1 - \frac{\ln(T/T_K)}{\sqrt{\ln^2(T/T_K) + \pi^2 S(S+1)}}\right], \tag{5.17}$$

where

$$\rho_0 = \frac{2mc_{\rm imp}}{\pi n e^2 \nu(0)} \tag{5.18}$$

corresponds to the maximum value possible for $s-$wave scattering. In other words, ρ_0 corresponds to the unitarity limit (see Eq.(4.27)) for the impurity scattering. At temperatures $T \gg T_K$, Eq.(5.17) will reduce to Eq.(5.12). At low temperatures ($T \ll T_K$), we would get

$$\rho_{\rm imp} \approx \rho_0 \left[1 - \frac{\pi^2}{4\ln^2(T/T_K)} + \cdots\right], \qquad T \ll T_K. \tag{5.19}$$

The resistivity is therefore predicted to saturate at low temperatures. We will not present details for the impurity susceptibility but the result of a similar analysis is that at low temperatures the expression we get for the susceptibility can be well approximated by

$$\chi_{\rm imp} \approx \frac{1}{1.22}\frac{(g\mu_B)^2}{4k_B}\frac{1}{T + 4.5T_K}, \qquad T \lesssim T_K. \tag{5.20}$$

The same analysis gives for the impurity specific heat,

$$C_{\text{imp}} \approx \frac{3\pi^2 k_B}{\ln^4(T/T_K)}, \qquad T \ll T_K. \tag{5.21}$$

Unfortunately, experiments reveal a completely different temperature dependence:

$$\rho_{\text{imp}} = \rho_0 \left[1 - \left(\frac{T}{\theta_\rho}\right)^2 + O\left(\frac{T}{\theta_\rho}\right)^4\right], \tag{5.22}$$

$$\chi_{\text{imp}}(T) = \chi_{\text{imp}}(0) \left[1 - \left(\frac{T}{\theta_\chi}\right)^2 + O\left(\frac{T}{\theta_\chi}\right)^4\right]. \tag{5.23}$$

Here the characteristic temperatures θ_ρ and θ_χ are of the order of the Kondo temperature. Such experimental observations suggest that the perturbative approach, regardless of how accurately we might do the perturbation theory, is unable to describe the physics below the Kondo temperature. The impurity resistivity and susceptibility resemble the corresponding behaviors we find in the resistivity and Pauli spin susceptibility of a Fermi liquid. Is there any similarity between the two? How should we describe the ground state and low temperature properties?

An answer was conjectured by Anderson and Yuval (1969-71) based on their scaling analysis of the Kondo model. They proposed that the effective coupling becomes infinitely large as the temperature approaches zero and the impurity forms a tightly bound singlet with a conduction electron. Once the impurity traps a conduction electron forming a tightly-bound singlet, the "impurity plus bound electron" complex behaves like a (non-magnetic) potential scatterer. There will possibly be some weak residual interactions of the complex with the conduction electrons, and, on the whole, the physics should resemble a weakly interacting Fermi gas.

5.2.2 Nozières' Fermi liquid theory for $T < T_K$

The correctness of Anderson and Yuval's conjecture was numerically confirmed by Wilson in 1976 using numerical renormalization group techniques. In the same year, Nozières assumed that Anderson's conjecture was correct and presented a phenomenological "Fermi liquid" model that enabled a theoretical derivation of the numerical results obtained by Wilson. We will discuss Nozières' analysis below.

When J is large, and the temperature is low, the impurity-conduction electron singlet that is formed cannot be broken by thermal means. The energy difference between the impurity singlet and triplet states is J. Nevertheless, *virtual* excitations of the singlet are possible because of the finite probability of electron hopping between the impurity and the neighboring sites. To appreciate this, let us begin by rewriting the Kondo Hamiltonian, Eq.(5.4), in real space,

$$H = -\sum_{<ij>;\alpha} t_{ij} c^{\dagger}_{i\alpha} c_{j\alpha} + J \sum_{\alpha\beta} \mathbf{S} \cdot \boldsymbol{\sigma}_{\alpha\beta} c^{\dagger}_{0\alpha} c_{0\beta}. \tag{5.24}$$

Here t_{ij} is the hopping element between sites i and j. We begin with the limit where $J \gg t$. If only nearest neighbor hopping is significant, then t_{ij} is related to $\epsilon_{\mathbf{k}}$ in Eq.(5.4) through $\epsilon_{\mathbf{k}} = -t \sum_{\mathbf{a}} \cos(\mathbf{k} \cdot \mathbf{a})$, where $\mathbf{a}$ are the lattice translation vectors. t_{ij} is of the order of the bandwidth, D. Let us examine the two ways in which the bound singlet at the impurity site can get excited to the triplet state due to the presence of the (kinetic) hopping term. We can either first move the bound electron out and then bring in a different electron with the opposite spin, or, we can first bring in an outside electron with the opposite spin and then take out the bound electron. Both these processes result in a finite overlap between the singlet and triplet impurity states, and the strength of this overlap is easily estimated to be $\sim t^2/J$ using second order perturbation theory. Furthermore, since the Pauli principle forbids the same site to have two electrons with the same spin, the electrons hopping between the impurity and the neighboring sites must have opposite spins. Since the impurity spin is oppositely oriented to the impurity electron (because we have a singlet), it follows that the hopping is possible only if electrons at the neighboring sites have the same spin as the impurity. Thus we have a *ferromagnetic* interaction induced between the impurity spin, and the neighboring electrons, e.g.,

$$\delta H_{\text{int}} \sim -\frac{t^2}{J} \sum_{\mathbf{a},\alpha\beta} c^{\dagger}_{\mathbf{a}\alpha} c_{\mathbf{a}\beta} \boldsymbol{\sigma}^{\alpha\beta} \cdot \mathbf{S}. \tag{5.25}$$

The conclusion is that an electron sitting at a site next to the impurity has a lower energy when it is oriented parallel to the impurity spin than the energy it would have if it were antiparallel to the impurity spin. This amounts to saying that if there were two electrons at the neighboring site (with opposite spins of course), then the one with its spin parallel to the impurity spin is attracted to the impurity while the other electron is repelled away. We thus have an effective *repulsion* between the two electrons at the neighboring site:

$$\delta H_{\text{int}} \sim \frac{t^4}{J^3} \sum_{\mathbf{a}} n_{\mathbf{a}\uparrow} n_{\mathbf{a}\downarrow}. \tag{5.26}$$

In this way two different conduction electrons interact with the same impurity and generate an effective interaction between themselves, mediated by the impurity. This is quite similar to superconductivity where an effective interaction is generated between two electrons when they scatter off the same phonon.

The impurity spin induces an effective repulsive interaction of the electrons. This system of interacting electrons can now be described using Fermi liquid theory. In a Fermi liquid, the state of the system is described by the distribution function $n_{\mathbf{p}}(\boldsymbol{\sigma})$ of the quasiparticle scattering states, and all physical quantities are functionals of $n_{\mathbf{p}}(\boldsymbol{\sigma})$. In particular, we believe this is true for the phase shifts $\delta_l(p, \boldsymbol{\sigma})$. For a point impurity, only $l = 0$ is relevant because all higher angular momentum electron wave functions will have a node at the origin. So when we write δ below, we actually mean only the $l = 0$ component.

We infer from Eq.(4.44) that if there is one bound state, i.e., $\Delta N = 1$, then $\delta(p, \sigma) = \pi/2$. At $T = 0$, this is indeed the case for our Kondo problem – there is precisely one bound electron forming a singlet with the impurity spin. At finite temperatures, we may expand the phase shift as

$$\delta(p, \boldsymbol{\sigma}) = \frac{\pi}{2} + \alpha(\epsilon_{\mathbf{p}}^{(0)} - \mu) + \mathrm{Tr}_{\boldsymbol{\sigma}'} \int (d\mathbf{p}')\phi(\mathbf{p}, \boldsymbol{\sigma}; \mathbf{p}', \boldsymbol{\sigma}')\delta n_{\mathbf{p}'}(\boldsymbol{\sigma}'), \qquad (5.27)$$

where δn as usual is the departure of the distribution from the ground state. We assume for simplicity that ϕ is independent of the momentum arguments. Since there is no change in the number of electrons in the system, we have

$$\mathrm{Tr}_{\boldsymbol{\sigma}'} \int (d\mathbf{p}')\delta n_{\mathbf{p}'}(\boldsymbol{\sigma}') = 0.$$

On the other hand, we expect a finite magnetization of the electrons when we ultimately apply a magnetic field, so let us say

$$\mu_B \mathrm{Tr}_{\boldsymbol{\sigma}'} \int (d\mathbf{p}')\boldsymbol{\sigma}'\delta n_{\mathbf{p}'}(\boldsymbol{\sigma}') = \mathbf{M} \neq 0.$$

If we are to write $\phi = \phi^s + (\boldsymbol{\sigma} \cdot \boldsymbol{\sigma}')\phi^a$, then the above arguments tell us that only ϕ^a appears in Eq.(5.27):

$$\delta(p, \boldsymbol{\sigma}) = \frac{\pi}{2} + \alpha(\epsilon_{\mathbf{p}}^{(0)} - \mu) + \phi^a \frac{\boldsymbol{\sigma} \cdot \mathbf{M}}{\mu_B}. \qquad (5.28)$$

The ϕ^a term would not have existed in the absence of the impurity spin.

At this point Nozières had a great insight that the "Fermi liquid" physics of the Kondo problem is essentially determined by quasiparticles close to the Fermi surface. In other words, the Kondo physics is tied to the Fermi surface. Therefore, he argued, if the chemical potential were changed by a small amount, the phase shifts should remain unchanged. Varying the *ground state* distribution with respect to μ while keeping every thing else fixed gives us the relation $\delta n_{\mathbf{p}'}(\boldsymbol{\sigma}') \approx -\delta(\epsilon_{\mathbf{p}}^{(0)} - \mu)\Delta\mu$. [1] We vary Eq.(5.27) with respect to the chemical potential and use $2\int(d\mathbf{p}')\delta(\epsilon_{\mathbf{p}}^{(0)} - \mu) = \nu(0)$ to get

$$\alpha + \phi^s \nu(0) = 0. \qquad (5.29)$$

[1] Note the negative sign. The reason for the negative sign is that if we kept the distribution function of the excited Fermi liquid fixed and increased the chemical potential of the system, then the number of excited particles has to decrease.

ϕ^s and ϕ^a are related in an interesting way. Eq.(5.26) tells us that the interaction is between antiparallel electrons. If we consider the general form of the interaction $\phi = \phi^s + (\boldsymbol{\sigma}\cdot\boldsymbol{\sigma}')\phi^a$, the antiparallel configuration, $\boldsymbol{\sigma}\cdot\boldsymbol{\sigma}' = -1$, has the interaction $\phi^s - \phi^a$, while the parallel configuration has the interaction $\phi^s + \phi^a$. We discussed this earlier at the end of Chapter 1.2, Eq.(1.31). As we only have interaction between antiparallel spins in our model, we must satisfy

$$\phi^s + \phi^a = 0. \tag{5.30}$$

We now estimate the impurity susceptibility and specific heat. The energy of the quasiparticles can be straightforwardly obtained from the phase shifts in the following manner. We first note that the average separation between the energy levels at the Fermi surface is $1/(\nu(0)V)$. Second, we have already learned that the number of bound states of each spin species is related to the corresponding phase shift through $\Delta N(\boldsymbol{\sigma}) = 2\delta(\boldsymbol{\sigma})/\pi$. The change in the quasiparticle density of states, $\Delta\nu$, is

$$\Delta\nu = \frac{1}{V}\frac{d\Delta N}{d\epsilon^{(0)}} = \frac{2}{\pi V}\frac{d\delta}{d\epsilon^{(0)}} = \frac{2\alpha}{\pi V}. \tag{5.31}$$

Likewise, the energies of the quasiparticles can be written as

$$\epsilon_{\mathbf{p}}(\boldsymbol{\sigma}) = \epsilon_{\mathbf{p}}^{(0)} - \mu_B \boldsymbol{\sigma}\cdot\mathbf{H} - \frac{2\delta(\boldsymbol{\sigma})}{\pi\nu(0)V} \tag{5.32}$$

$$= \epsilon_{\mathbf{p}}^{(0)} + \mu_B \boldsymbol{\sigma}\cdot\mathbf{B}_{\text{ext}} - \frac{2}{\pi\nu(0)V}\left[\alpha(\epsilon_{\mathbf{p}}^{(0)} - \mu) + \phi^a \frac{\boldsymbol{\sigma}\cdot\mathbf{M}}{\mu_B}\right]. \tag{5.33}$$

$$= \epsilon_{\mathbf{p}}^{(0)} - \mu_B \boldsymbol{\sigma}\cdot\mathbf{B}_{\text{ext}} - \frac{2\alpha}{\pi\nu(0)V}\left[(\epsilon_{\mathbf{p}}^{(0)} - \mu) + \frac{\boldsymbol{\sigma}\cdot\mathbf{M}}{\nu(0)\mu_B}\right], \tag{5.34}$$

where we used $\alpha = -\phi^s\nu(0) = \phi^a\nu(0)$ in the second equation. If we use $\mathbf{M} = \chi_{\text{tot}}\mathbf{B}_{\text{ext}}$ and rearrange the above equation, we can say

$$\epsilon_{\mathbf{p}}(\boldsymbol{\sigma}) - \mu = (\epsilon_{\mathbf{p}}^{(0)} - \mu)\left[1 - \frac{2\alpha}{\pi\nu(0)V}\right] - \mu_B\boldsymbol{\sigma}\cdot\mathbf{B}_{\text{ext}}\left[1 + \frac{2\alpha}{\pi\nu(0)V}\frac{\chi_{\text{tot}}}{\nu(0)\mu_B^2}\right] \tag{5.35}$$

$$= (\epsilon_{\mathbf{p}}^{(0)} - \mu)(1 - \Delta\nu/\nu(0)) - \mu_B\boldsymbol{\sigma}\cdot\mathbf{B}_{\text{ext}}(1 + \Delta\nu/\nu(0)), \tag{5.36}$$

where in the second equation we used $\chi_{\text{tot}} = \nu(0)\mu_B^2 + O(\alpha)$. This leads us to the following Pauli susceptibility,

$$\chi_{\text{tot}} = \chi + \chi_{\text{imp}} = \mu_B^2\nu(0)\frac{1 + \Delta\nu/\nu(0)}{1 - \Delta\nu/\nu(0)} \approx \chi(1 + 2\Delta\nu/\nu(0)). \tag{5.37}$$

The impurity contribution to the Pauli susceptibility is therefore

$$\chi_{\text{imp}} = \frac{2\Delta\nu}{\nu(0)}\chi. \tag{5.38}$$

The change density of states leads to a change in the specific heat per impurity:

$$C_{\text{imp}} = \Delta\nu \left(\frac{\pi^2 k_B^2}{3}\right) T = \frac{\Delta\nu}{\nu(0)} C_V. \tag{5.39}$$

Combining Eq.(5.38) and Eq.(5.39) we obtain

$$R = \frac{(\chi_{\text{imp}}/\chi)}{(C_{\text{imp}}/C_V)} = 2. \tag{5.40}$$

R is known as the Wilson-Sommerfeld ratio. Wilson first obtained this value numerically but it was Nozières who showed that $R = 2$ exactly. This is a remarkable result. Somehow the impurity affects the susceptibility twice as much as it affects the specific heat when compared with the corresponding result for a Fermi gas. $R = 2$ is used as a check in experiments for a confirmation of the Kondo effect. Note that this universal value would not be obtained in a general Fermi liquid.

Nozières could also obtain the correct temperature dependence of the impurity contribution to the resistivity as well as other properties we presented earlier in this chapter. The details of the calculations may be looked up in the original paper.[2]

[2]P. Nozières, *Journal of Low Temperature Physics* **17**, 31 (1974).

Part II

Microscopic Approach

Chapter 6

Response of an Ideal Fermi Gas

6.1 Thomas-Fermi and Lindhard polarizability

We studied in Sec. 1.3 two spatially uniform response functions of a Fermi liquid, namely the compressibility and Pauli susceptibility. Often the perturbation is spatially (and temporally) nonuniform and the response of the Fermi liquid cannot in general be described by uniform susceptibilities. For such cases we need to know the momentum (and frequency) dependence of the susceptibility.

We could proceed with this plan straightaway for our familiar neutral Fermi liquid; but perhaps it is more useful to study a more commonly encountered situation where the fermions are charged, such as electrons. Because of the long-range nature of the Coulomb interaction of the electrons, a given electron will interact with a large number of other electrons, and not just its neighbors. This raises the question as to whether we can use the Fermi liquid formalism that we have developed for short-range interactions. Fortunately, it turns out that the *effective* interaction between the electrons is not long-range in many circumstances, and, after some modifications, Fermi liquid theory may still be used. The effective interaction between two electrons can be short-range because the other electrons redistribute themselves in an attempt to compensate, or screen, the electric fields of the two electrons. A high density of electrons increases this screening effect. If the electron density is not high, the screening will not be very effective, and the range of interaction will be large. In that case a Fermi liquid description will not be appropriate.

The electron density above which the Fermi liquid description will be valid may be estimated by the following simple argument. If the average kinetic energy per unit volume of the electrons is much larger than the average potential energy of the electrons per unit volume, the effects of Coulomb interaction can be expected to be small. The kinetic energy density of the electron gas is $K = (3/5)n\epsilon_F \simeq (\hbar^2/2m)n^{5/3}$. The potential energy density may be expressed in terms of the inter-electron distance r_{ee} as $U = n(e^2/4\pi\epsilon_0 r_{ee}) \simeq (e^2/\epsilon_0)n^{4/3}$. Here we have used $\epsilon_F = \hbar^2 k_F^2/2m$ and $1/r_{ee} \simeq n^{1/3}$. The condition $K \gg U$ implies

$$n^{1/3} \times \left(\frac{4\pi\epsilon_0\hbar^2}{me^2}\right) = \frac{a_B}{r_{ee}} \gg 1, \tag{6.1}$$

where a_B is the Bohr radius. Thus the condition for the validity of Landau Fermi liquid theory for the case of electrons is that their average separation is much less than the Bohr radius.

Let us now calculate the wavevector dependent electrical response of such a dense electron gas. For simplicity, we assume Coulomb interaction is the only interaction between the electrons.

6.1.1 Polarizability and dielectric function

First we need to get acquainted with some definitions. When an external charge with density $\rho_{\text{ext}}(\mathbf{r})$ and a bare potential $\phi^{(0)}(\mathbf{r})$ is introduced into an electron gas, the redistribution of the electrons gives rise to an induced charge density $\rho_{\text{ind}}(\mathbf{r})$ and induced potential $\phi_{\text{int}}(\mathbf{r})$. The total charge density $\rho(\mathbf{r}) = \rho_{\text{ext}}(\mathbf{r}) + \rho_{\text{ind}}(\mathbf{r})$ and the total (effective) potential $\phi(\mathbf{r}) = \phi^{(0)}(\mathbf{r}) + \phi_{\text{ind}}(\mathbf{r})$ will satisfy the Poisson's equation:

$$-\nabla^2\phi(\mathbf{r}) = \rho(\mathbf{r})/\epsilon_0. \tag{6.2}$$

The bare potential and induced potential respectively obey

$$-\nabla^2\phi^{(0)}(\mathbf{r}) = \rho_{\text{ext}}(\mathbf{r})/\epsilon_0, \text{ and}$$
$$-\nabla^2\phi_{\text{ind}}(\mathbf{r}) = \rho_{\text{ind}}(\mathbf{r})/\epsilon_0. \tag{6.3}$$

We may also write the Poisson equations in the momentum representation; for example, $k^2\phi_{\text{ind}}(\mathbf{k}) = \rho_{\text{ind}}(\mathbf{k})/\epsilon_0$. The dielectric function $\varepsilon(\mathbf{k})$ is the ratio of the *bare* potential to the *total* potential:

$$\phi(\mathbf{k}) = \frac{\phi^{(0)}(\mathbf{k})}{\varepsilon(\mathbf{k})}. \tag{6.4}$$

If $\varepsilon(\mathbf{k})$ is independent of the external charge density ρ_{ext}, the total potential is directly proportional to ρ_{ext}; such screening is described as *linear.* This is not always the case, but, in this chapter, we will only be studying linear screening. Another useful quantity is the polarizability, $\Pi(\mathbf{k})$, which is the ratio of the induced charge density to the total potential:

$$\rho_{\text{ind}}(\mathbf{k}) = \Pi(\mathbf{k})\phi(\mathbf{k}). \tag{6.5}$$

In linear screening, the polarizability is independent of the total potential (which depends on the externally introduced charge) and is simply an intrinsic property of the system.Using this definition of the polarizability, we may express the total (effective) potential as

$$\phi(\mathbf{k}) = \frac{\phi^{(0)}(\mathbf{k})}{1 - \frac{1}{\epsilon_0 \mathbf{k}^2}\Pi(\mathbf{k})} = \frac{\phi^{(0)}(\mathbf{k})}{\varepsilon(\mathbf{k})}. \tag{6.6}$$

$V(\mathbf{k})$ is also known as the screened Coulomb potential since it contains the effects of the redistribution of the electronic charge. The susceptibility, $\chi(\mathbf{k})$, is related to the dielectric function through $\varepsilon(\mathbf{k}) = 1 + \chi(\mathbf{k})$.

6.1.2 Thomas-Fermi polarizability

If the characteristic length scale L over which the total potential $\phi(\mathbf{r})$ varies is much larger than k_F^{-1}, then the response of the electron gas may be described semiclassically. In the semiclassical approximation, the total energy $\epsilon_{\mathbf{k}}(\mathbf{r})$ of an electron with wavevector $\mathbf{k}$ may be written as

$$\epsilon_{\mathbf{k}}(\mathbf{r}) = \epsilon_{\mathbf{k}}^{(0)} - e\phi(\mathbf{r}). \tag{6.7}$$

The induced charge density is

$$\rho_{\text{ind}}(\mathbf{r}) = -\frac{2e}{V}\sum_{\mathbf{k}}[n_{\mathbf{k}}(\mathbf{r}) - n_{\mathbf{k}}^{(0)}], \tag{6.8}$$

where $n_{\mathbf{k}}(\mathbf{r})$ is the Fermi-Dirac distribution function corresponding to the energy $\epsilon_{\mathbf{k}}(\mathbf{r})$. If $e\phi(\mathbf{r})$ is small compared to the Fermi energy, then we can easily work out the polarizability:

$$\begin{aligned} \rho_{\text{ind}}(\mathbf{r}) &\approx 2e^2\phi(\mathbf{r}) \int (d\mathbf{k}) \frac{\partial n_{\mathbf{k}}^{(0)}}{\partial \epsilon_{\mathbf{k}}^{(0)}} \\ &= -e^2\nu(0)\phi(\mathbf{r}) \equiv \Pi\phi(\mathbf{r}). \end{aligned} \tag{6.9}$$

Using the polarizability obtained in Eq.(6.9) in the expression for the effective potential, Eq.(6.6), we have

$$\phi(\mathbf{k}) = \frac{\phi^{(0)}(\mathbf{k})}{1 + \frac{e^2\nu(0)}{\epsilon_0 \mathbf{k}^2}}. \tag{6.10}$$

If $\phi^{(0)}$ is due to a point unit-charge, i.e., $\rho_{\text{ext}}(\mathbf{r}) = e\delta(\mathbf{r})$, then it is easy to see from the first of Eqs.(6.3) that $\phi^{(0)}(\mathbf{k}) = e/\epsilon_0\mathbf{k}^2$. Thus the $\mathbf{k}-$dependence of the effective potential of the point charge is

$$\phi(\mathbf{k}) = \frac{e/\epsilon_0}{\mathbf{k}^2 + (e^2\nu(0)/\epsilon_0)}. \tag{6.11}$$

In real space, the effective potential decreases exponentially at large distances,

$$\phi(\mathbf{r}) = \frac{e}{4\pi\epsilon_0 r} \exp[-r/\lambda_{TF}]. \tag{6.12}$$

The length scale,

$$\lambda_{TF} = \sqrt{\epsilon_0/\nu(0)e^2}, \tag{6.13}$$

over which the effective potential varies appreciably, is known as the Thomas-Fermi screening length. For the validity of our semiclassical treatment, we require $\lambda_{TF}k_F \gg 1$. There is an added justification for this requirement. In Eq.(6.1) we found that the Coulomb interaction energy can be considered small in comparison with the kinetic energy when $a_B/r_{ee} \gg 1$ is satisfied. This condition is also equivalent to $\lambda_{TF}k_F \gg 1$.

A quick way to obtain Eq.(6.11) is to consider the Poisson equation for the total potential, Eq.(6.2), with $\rho = e\delta(\mathbf{r}) + \rho_{\text{ind}}(\mathbf{r})$, and use Eq.(6.8) for the induced charge density. Taking the Fourier transform on both sides of the Poisson equation, and rearranging, we get the solution, Eq.(6.11).

The disadvantage of the Thomas-Fermi treatment of screening is that its validity is limited to the semiclassical regime. Its advantage is that the treatment still remains valid even when ρ_{ind} is not linearly dependent on $\phi(\mathbf{r})$. Such a situation appears, for instance, in many disordered semiconductors. The most well-known example is perhaps the Coulomb gap in disordered semiconductors discussed by Efros and Shklovskii, where the induced charge density is proportional to $\phi^3(\mathbf{r})$.

6.1.3 Lindhard polarizability

The semiclassical approximation, necessary in the Thomas-Fermi approach, is not valid if the potential changes appreciably over a length scale of the order of k_F^{-1}. The Lindhard approach we will study now is valid at all wavevectors. However, unlike Thomas-Fermi, it requires that the electrical response is linear.

We begin with the Schrödinger equation for an electron moving in the effective potential $\phi(\mathbf{r})$,

$$\left[-\frac{\hbar^2}{2m}\nabla^2 - e\phi(\mathbf{r})\right]\psi(\mathbf{r}) = \epsilon\psi(\mathbf{r}), \tag{6.14}$$

and treat $\phi(\mathbf{r})$ as a perturbation. The wavefunction ψ and the energy E may be expanded in increasing powers of $\phi(\mathbf{r})$. We denote this series as $\psi = \psi^{(0)} + \psi^{(1)} + \cdots$, and $\epsilon = \epsilon^{(0)} + \epsilon^{(1)} + \cdots$, where the numbers in the superscripts represent the orders of expansion in $\phi(\mathbf{r})$. Thus $\psi_{\mathbf{k}}^{(0)}(\mathbf{r}) = (1/\sqrt{V})e^{i\mathbf{k}\cdot\mathbf{r}}$ and $\epsilon_{\mathbf{k}}^{(0)} = \hbar^2\mathbf{k}^2/2m$. The first order correction to the unperturbed wavefunction is

$$\psi_{\mathbf{p}}^{(1)}(\mathbf{r}) = -e\sum_{\mathbf{k};\mathbf{k}\neq\mathbf{p}}\frac{\phi(\mathbf{k}-\mathbf{p})}{\epsilon_{\mathbf{p}}^{(0)} - \epsilon_{\mathbf{k}}^{(0)}}\psi_{\mathbf{k}}^{(0)}(\mathbf{r}), \tag{6.15}$$

where $\phi(\mathbf{q}) = (1/V)\int d\mathbf{r}\,\phi(\mathbf{r})e^{-i\mathbf{q}\cdot\mathbf{r}}$. For simplicity, we assume $\phi(\mathbf{q}=0) = 0$, i.e., the effective potential is not constant. As $\phi(\mathbf{q}=0)$ is the first order correction to the unperturbed energy levels, our assumption means that the energy levels are unchanged to first order in the perturbation.

The induced charge density is given by

$$\begin{aligned}
\rho_{\text{ind}}(\mathbf{r}) &= -2e\sum_{\mathbf{k}} n_{\mathbf{k}}^{(0)}[\psi_{\mathbf{k}}^*\psi_{\mathbf{k}} - (\psi_{\mathbf{k}}^{(0)})^*\psi_{\mathbf{k}}^{(0)}] \\
&\approx -2e\sum_{\mathbf{k}} n_{\mathbf{k}}^{(0)}[(\psi_{\mathbf{k}}^{(0)})^*\psi_{\mathbf{k}}^{(1)} + \text{c.c.}] \\
&= \frac{2e^2}{V}\sum_{\mathbf{k}}\sum_{\mathbf{k}';\mathbf{k}'\neq\mathbf{k}} n_{\mathbf{k}}^{(0)}\left[\frac{\phi(\mathbf{k}'-\mathbf{k})}{\epsilon_{\mathbf{k}}^{(0)} - \epsilon_{\mathbf{k}'}^{(0)}}e^{i(\mathbf{k}'-\mathbf{k})\cdot\mathbf{r}} + \frac{\phi^*(\mathbf{k}'-\mathbf{k})}{\epsilon_{\mathbf{k}}^{(0)} - \epsilon_{\mathbf{k}'}^{(0)}}e^{-i(\mathbf{k}'-\mathbf{k})\cdot\mathbf{r}}\right].
\end{aligned} \tag{6.16}$$

Introducing $\mathbf{q} = \mathbf{k}' - \mathbf{k}$ and using $\phi^*(\mathbf{q}) = \phi(-\mathbf{q})$ we can re-express Eq.(6.16) as

$$\begin{aligned}
\rho_{\text{ind}}(\mathbf{r}) &= \frac{2e^2}{V}\sum_{\mathbf{q}\neq 0}\sum_{\mathbf{k}} n_{\mathbf{k}}^{(0)}\left[\frac{\phi(\mathbf{q})}{\epsilon_{\mathbf{k}}^{(0)} - \epsilon_{\mathbf{k}+\mathbf{q}}^{(0)}}e^{i\mathbf{q}\cdot\mathbf{r}} + \frac{\phi(-\mathbf{q})}{\epsilon_{\mathbf{k}}^{(0)} - \epsilon_{\mathbf{k}+\mathbf{q}}^{(0)}}e^{-i\mathbf{q}\cdot\mathbf{r}}\right] \\
&= \frac{2e^2}{V}\sum_{\mathbf{q}\neq 0}\sum_{\mathbf{k}} n_{\mathbf{k}}^{(0)}\left[\frac{1}{\epsilon_{\mathbf{k}}^{(0)} - \epsilon_{\mathbf{k}+\mathbf{q}}^{(0)}} + \frac{1}{\epsilon_{\mathbf{k}}^{(0)} - \epsilon_{\mathbf{k}-\mathbf{q}}^{(0)}}\right]\phi(\mathbf{q})e^{i\mathbf{q}\cdot\mathbf{r}}.
\end{aligned}$$

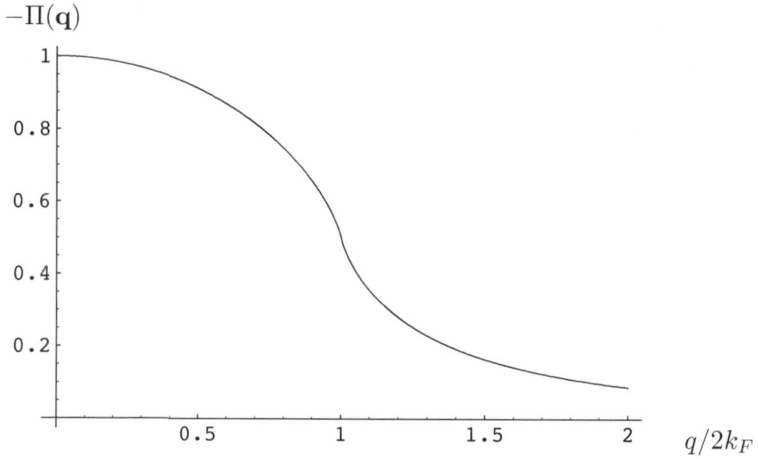

FIGURE 6.1: A plot of the Lindhard polarizability $\Pi(\mathbf{q})$ in three dimensions as given by Eq.(6.19) measured in units of the Thomas-Fermi value. $\Pi(\mathbf{q})$ is weakly singular at $q = 2k_F$.

We can now read off the Lindhard polarizability ($\rho_{\text{ind}}(\mathbf{q}) = \Pi(\mathbf{q})\phi(\mathbf{q})$):

$$\Pi(\mathbf{q}) = 2e^2 \int (d\mathbf{k})\, n_{\mathbf{k}}^{(0)} \left[\frac{1}{\epsilon_{\mathbf{k}}^{(0)} - \epsilon_{\mathbf{k}+\mathbf{q}}^{(0)}} + \frac{1}{\epsilon_{\mathbf{k}}^{(0)} - \epsilon_{\mathbf{k}-\mathbf{q}}^{(0)}} \right]$$

$$= 2e^2 \int (d\mathbf{k}) \frac{n_{\mathbf{k}}^{(0)} - n_{\mathbf{k}+\mathbf{q}}^{(0)}}{\epsilon_{\mathbf{k}}^{(0)} - \epsilon_{\mathbf{k}+\mathbf{q}}^{(0)}} \equiv 2e^2 \int (d\mathbf{k}) \frac{n_{\mathbf{k}-\mathbf{q}/2}^{(0)} - n_{\mathbf{k}+\mathbf{q}/2}^{(0)}}{\epsilon_{\mathbf{k}-\mathbf{q}/2}^{(0)} - \epsilon_{\mathbf{k}+\mathbf{q}/2}^{(0)}}. \tag{6.17}$$

In the limit $\mathbf{q} \to 0$, the Lindhard polarizability approaches the Thomas-Fermi value:

$$\Pi(\mathbf{q})\Big|_{\mathbf{q}\to 0} = 2e^2 \int (d\mathbf{k}) \frac{\partial n_{\mathbf{k}}^{(0)}}{\partial \epsilon_{\mathbf{k}}^{(0)}} = -e^2 \nu(0). \tag{6.18}$$

The calculation of $\Pi(\mathbf{q})$ simplifies at $T = 0$ because the distribution function becomes a unit step function, $n_{\mathbf{k}}^{(0)} = \Theta(\mu - \epsilon_{\mathbf{k}}^{(0)})$. For a general value of $\mathbf{q}$, it is convenient to evaluate the Lindhard polarizability using the first equation in Eq.(6.17). For example, in three dimensions we have

$$\Pi(\mathbf{q}) = -\left(\frac{m}{\hbar^2 \pi^2 q}\right) e^2 \int_0^{k_F} dk\, k^2 \int_0^{\pi} d\theta\, \sin\theta \left[\frac{1}{q + 2k\cos\theta} + \frac{1}{q - 2k\cos\theta} \right]$$

$$= -e^2 \nu(0) \left[\frac{1}{2} + \frac{1 - (q/2k_F)^2}{4(q/2k_F)} \ln \left| \frac{1 + (q/2k_F)}{1 - (q/2k_F)} \right| \right]. \tag{6.19}$$

Figure 6.1 shows a plot of the Lindhard polarizability in three dimensions, measured in units of the Thomas-Fermi value. There is a weak singularity at $q = 2k_F$. At very large wavevectors $q \gg 2k_F$, the polarizability approaches zero, signifying that the electrons are unable to screen the external charge as effectively as they do near $q = 0$.

In two dimensions we get an interesting result,

$$\Pi(\mathbf{q}) = -e^2 \left(\frac{m}{\pi\hbar^2}\right) \begin{cases} 1, & q/2k_F < 1 \\ 1 - \sqrt{1-(2k_F/q)^2}, & q/2k_F \geq 1. \end{cases} \tag{6.20}$$

The two-dimensional Lindhard polarizability is constant and equal to the Thomas-Fermi value up to $q = 2k_F$. Beyond $q = 2k_F$, it decreases in a square-root manner.

We have not yet discussed the effective potential $\phi(\mathbf{r})$ of a unit point charge immersed in the electron gas when the polarizability is given by the Lindhard expression. That will be the subject of Sec. 6.2. Here we would just state that the result is nothing like the Thomas-Fermi expression, Eq.(6.12), and instead decreases much more slowly with distance, undergoing oscillations.

6.1.3.1 Frequency dependence

At finite frequencies, the Lindhard polarizability can be shown to be

$$\Pi(\mathbf{q}, \omega) = 2e^2 \int (d\mathbf{k}) \frac{n_{\mathbf{k}}^{(0)} - n_{\mathbf{k}+\mathbf{q}}^{(0)}}{\epsilon_{\mathbf{k}}^{(0)} - \epsilon_{\mathbf{k}+\mathbf{q}}^{(0)} + \hbar\omega + i\delta}. \tag{6.21}$$

6.2 Friedel and RKKY oscillations

The Kondo effect we studied earlier is an example of a local perturbation of a Fermi gas. The impurity spin is the local perturbation. Although we were primarily interested in obtaining the impurity properties, we also found that the presence of the impurity spin affected the properties of the electron gas. In particular, the impurity spin generated an effective interaction of the conduction electrons the strength of which decreased rapidly with the distance from the impurity. In this section we will study how an electron gas responds to various kinds of *local* perturbations such as potential scatterers and magnetic impurities.

For the potential scatterer as well as the magnetic impurity, the response of the electron gas shows characteristic oscillations (in real space) with a period of π/k_F, and a power-law decay. For the case of potential scattering, they are known as Friedel oscillations, and for a magnetic impurity, they are known as Ruderman-Kittel-Kasuya-Yosida (RKKY) oscillations.

6.2.1 Friedel oscillations

One way of obtaining the spatial dependence of the induced charge density $\rho_{\text{ind}}(\mathbf{r})$ due to a point external charge (say of unit size) is by taking the Fourier transform of $\rho_{\text{ind}}(\mathbf{q})$,

$$\rho_{\text{ind}}(\mathbf{r}) = \int (d\mathbf{q})\rho_{\text{ind}}(\mathbf{q})e^{i\mathbf{q}\cdot\mathbf{r}}.$$

We showed in Sec. 6.1 that $\rho_{\text{ind}}(\mathbf{q}) = \Pi(\mathbf{q})\phi(\mathbf{q})$, where

$$\Pi(\mathbf{q}) = -e^2\nu(0)\left[\frac{1}{2} + \frac{1-(q/2k_F)^2}{4(q/2k_F)}\ln\left|\frac{1+(q/2k_F)}{1-(q/2k_F)}\right|\right]$$

is the Lindhard polarizability, and

$$\phi(\mathbf{q}) = \frac{(e/q^2\epsilon_0)}{1-\frac{1}{q^2\epsilon_0}\Pi(\mathbf{q})}$$

is the screened Coulomb potential of the unit point external charge. $\rho_{\text{ind}}(\mathbf{r})$ has a simple-looking form in terms of the dielectric function:

$$\rho_{\text{ind}}(\mathbf{r}) = -e\int (d\mathbf{q})\left(1-\frac{1}{\varepsilon(\mathbf{q})}\right)e^{i\mathbf{q}\cdot\mathbf{r}}. \tag{6.22}$$

One immediate result is that the total induced charge has the same magnitude and opposite sign as the external charge,

$$\begin{aligned}\int d\mathbf{r}\,\rho_{\text{ind}}(\mathbf{r}) &= -e\int \frac{d\mathbf{q}}{(2\pi)^3}\left(1-\frac{1}{\varepsilon(\mathbf{q})}\right)\int d\mathbf{r}\,e^{i\mathbf{q}\cdot\mathbf{r}} \\ &= -e\int d\mathbf{q}\left(1-\frac{1}{\varepsilon(\mathbf{q})}\right)\delta(\mathbf{q}) = -e.\end{aligned} \tag{6.23}$$

Here we made use of the fact that $\varepsilon(\mathbf{q}) = \infty$ at $\mathbf{q} = 0$. Figure 6.2 shows the result of a numerical evaluation of the Fourier transform in Eq.(6.22); the induced charge density fits asymptotically to

$$\rho_{\text{ind}}(\mathbf{r}) \sim \frac{e}{r^3}\cos(2k_F r), \qquad k_F r \gg 1. \tag{6.24}$$

In comparison, the Thomas-Fermi screened Coulomb potential would have given us an induced charge density proportional to $-(1/r)e^{-r/\lambda_{TF}}$. (see Sec. 6.1.2)

It is possible to obtain the Friedel oscillations, Eq.(6.24), in a much simpler and transparent manner using scattering theory. For our scattering treatment to be valid, we require the total potential $\phi(\mathbf{r})$ due to an external point charge placed in the medium to be "short-range" in the sense that $[e\int d\mathbf{r}\phi(\mathbf{r})]^2$ converges. If $\phi(\mathbf{r})$ varies as $e\cos(2k_F r)/(\epsilon_0 k_F^2 r^3)$ and the lower limit for r is chosen

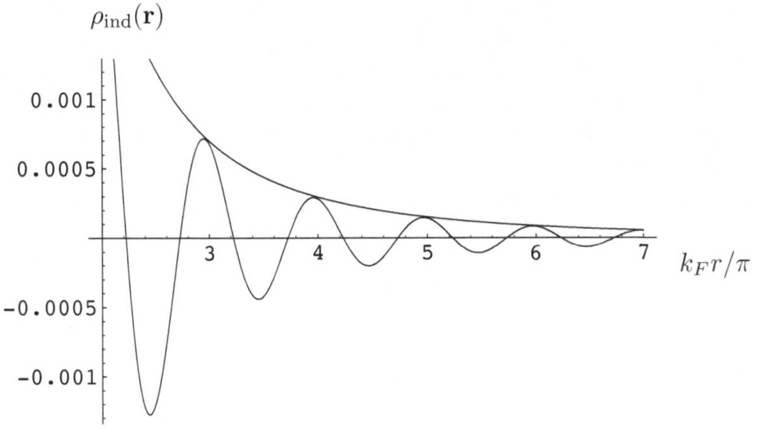

FIGURE 6.2: The oscillatory curve is the numerically evaluated charge density $\rho_{\rm ind}(\mathbf{r})$ induced by an impurity charge placed at the origin using Eq.(6.22) and the Lindhard dielectric function. The $x-$axis shows the distance from the impurity measured in units of π/k_F. The envelope of the oscillatory curve is proportional to $1/r^3$. The calculations are for a particular density of conduction electrons such that $k_F a_B = 1/\pi$. The induced charge density is seen to be asymptotically proportional to $\cos(2k_F r)/r^3$. These oscillations of the electron charge density induced by an external point charge are known as Friedel oscillations.

to be the lattice size $a \sim 1/k_F$, we estimate $|e \int d\mathbf{r}\phi(\mathbf{r})| < (e^2/\epsilon_0 k_F^2)\ln(2k_F a)$ which is convergent.

In Sec. 4, Eq.(4.21), we learned that the wavefunction behaves as

$$\begin{aligned}\psi_{\mathbf{k}}(\mathbf{r}) =& \frac{1}{2}\sum_{l=0}^{\infty} i^l(2l+1)[h_l^{(2)}(kr) + e^{i2\delta_l(k)}h_l^{(1)}(kr)] \\ \simeq& \frac{1}{i2kr}\sum_{l=0}^{\infty}(2l+1)[-e^{-ikr} + e^{i2\delta_l(k)}e^{ikr}] \\ =& \frac{1}{kr}\sum_{l=0}^{\infty}(2l+1)\sin(kr+\delta_l(k)),\end{aligned} \tag{6.25}$$

at large distances from the scatterer such that $kr \gg l\pi/2$. We used $h_l^{(1)} \simeq i^{-l}e^{ikr}/ikr$, and $h_l^{(2)} = (h_l^{(1)})^*$. The unperturbed wavefunction $\psi_{\mathbf{k}}^{(0)}$ corresponds to $\delta_l = 0$. The induced charge density is

$$\begin{aligned}\rho_{\text{ind}}(\mathbf{r}) &= -2e\int_0^{\infty}(d\mathbf{k})n_{\mathbf{k}}^{(0)}[|\psi_{\mathbf{k}}(\mathbf{r})|^2 - |\psi_{\mathbf{k}}^{(0)}(\mathbf{r})|^2] \\ &= -\frac{e}{4\pi^3 r^2}\sum_{l=0}^{\infty}(2l+1)\int_0^{k_F} dk\,[\sin^2(kr+\delta_l(k)) - \sin^2(kr)] \\ &= -\frac{e}{4\pi^3 r^2}\sum_{l=0}^{\infty}(2l+1)\int_0^{k_F} dk\,[\cos(2kr) - \cos(2kr+2\delta_l(k))].\end{aligned} \tag{6.26}$$

Integration over the wavevectors would have been trivial were it not for the fact that the phase shifts δ_l are themselves functions of k. We can simplify the problem a bit by arguing that elastic scattering by the external charge (which is the case we are studying) is likely to be weak for wavevectors k lying much below the Fermi surface because the Fermi sea is filled up. Therefore we expand $\delta_l(k)$ in a power series near the Fermi wavevector,

$$\delta_l(k) = \delta_l(k_F) + b_l(k-k_F) + \cdots. \tag{6.27}$$

With the above approximation, Eq.(6.26) simplifies to

$$\begin{aligned}\rho_{\text{ind}}(\mathbf{r}) \simeq& -\frac{e}{4\pi^3 r^2}\sum_{l=0}^{\infty}(2l+1) \\ &\times \int_0^{k_F} dk\ (\cos(2kr) - \cos[2k(r+b_l) + \delta_l(k_F) - 2k_F b_l]) \\ \approx& -\frac{e}{4\pi^3 r^3}\sum_{l=0}^{\infty}(2l+1)\cos[2k_F r + \delta_l(k_F)]\sin[\delta_l(k_F)].\end{aligned} \tag{6.28}$$

If the phase shifts are small, we can approximate Eq.(6.28) as

$$\rho_{\text{ind}}(\mathbf{r}) \approx -\frac{e\cos(2k_F r)}{4\pi^3 r^3}\sum_{l=0}^{\infty}(2l+1)\delta_l(k_F). \tag{6.29}$$

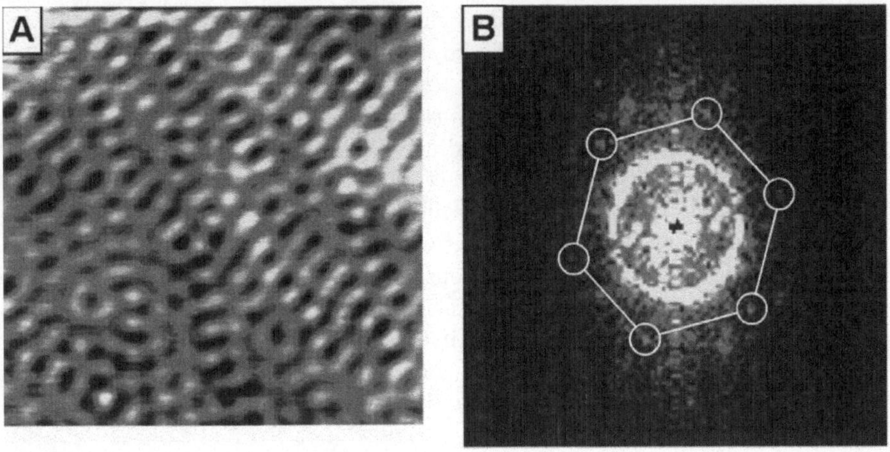

FIGURE 6.3: (a) STM picture of Friedel oscillations caused by impurities on a beryllium surface. (b) The 2D Fourier transform of the STM image. The hexagon indicates the reciprocal lattice and the $2k_F$ ring is due to Friedel oscillations. Source: P. T. Sprunger *et al.*, *Science* **275**, 1764 (1997).

We recognize the sum in Eq.(6.29) as $\pi\Delta N/2$ (Friedel sum rule), where ΔN is the number of electrons bound to the external positive charge of unit magnitude we have introduced. The total number of electrons bound to the external charge is one, and this is also the screening charge. With this understanding, the induced charge density can be approximately written as

$$\rho_{\text{ind}}(\mathbf{r}) \approx -\frac{e\cos(2k_F r)}{8\pi^2 r^3}. \tag{6.30}$$

Although Eq.(6.30) has a very appealing form, Eq.(6.28) should be regarded as the more correct result.

The long-range density oscillations, although small, have nevertheless been detected in nuclear magnetic resonance experiments.[1] More recently, there have been many reports of the direct observation, using scanning tunneling microscopy (STM) techniques, of Friedel oscillations in conductors.[2]

Figure 6.3 shows an STM picture of Friedel oscillations on a beryllium surface with impurities.

6.2.2 RKKY interaction

Magnetic impurities in conductors give rise to long range $2k_F$ ripples of the magnetization of the electrons similar to the Friedel oscillations of electron

[1] T. Rowland, *Phys. Rev.* **119**, 900 (1960), W. Kohn and S. Vosko, *Phys. Rev.* **119**, 912 (1960).

[2] See for example P. T. Sprunger *et al.*, *Science* **275**, 1764 (1997).

charge density that arise when the electron gas is perturbed locally. Suppose one places another magnetic impurity at some other point in the conductor. Then depending on whether the second impurity is located at a magnetization crest or trough of the ripples caused by the first impurity, the spin orientation of the second impurity will adjust accordingly to minimize the energy. By changing the separation of the two impurities, one then goes through a succession of ferromagnetic and antiferromagnetic couplings between the impurities. Such an oscillatory indirect exchange interaction between the impurities, mediated by conduction electrons, is known as the Ruderman-Kittel-Kasuya-Yosida or RKKY interaction. The story of the RKKY interaction dates back to 1954 when Ruderman and Kittel studied[3] the effective exchange interaction between two nuclear spins that couple to the conduction electrons through the hyperfine interaction. Kasuya and Yosida generalized the treatment to include magnetic impurities arising from unfilled d or f orbitals of the impurity atom embedded in a metallic host. We briefly discuss the analysis of Ruderman and Kittel.

Consider the Schrödinger equation of a conduction electron in a lattice interacting with impurity spins $\mathbf{I}_i$ at certain sites $\mathbf{r}_i$ through an exchange coupling:

$$\left[-\frac{\hbar^2}{2m}\nabla^2 + U(\mathbf{r}) + \sum_i J(\mathbf{r}-\mathbf{r}_i)\mathbf{I}_i\cdot\boldsymbol{\sigma}\right]\psi(\mathbf{r}) = E\psi(\mathbf{r}). \tag{6.31}$$

Here $U(\mathbf{r})$ is the periodic potential of the lattice, and $\boldsymbol{\sigma}$ are Pauli matrices. We assume that the conduction electron coupling to the impurities is weak, $J/\epsilon_F \ll 1$. In the absence of the impurities, the solutions of the Schrödinger equation are Bloch waves, $\psi_{\mathbf{k}}(\mathbf{r}) = e^{i\mathbf{k}\cdot\mathbf{r}}u_{\mathbf{k}}(\mathbf{r})$, which we choose to normalize as

$$\frac{1}{V}\int d\mathbf{r}\psi^*_{\mathbf{k}}(\mathbf{r})\psi_{\mathbf{k}}(\mathbf{r}) = 1.$$

This way, the Bloch functions do not have any dimension. Consider the matrix element $V^{(i)}(\mathbf{k},\mathbf{k}')$ for the interaction with the ith impurity which is given by

$$\begin{aligned}
V^{(i)}(\mathbf{k},\mathbf{k}') &= \mathbf{I}_i\cdot\boldsymbol{\sigma}\int d\mathbf{r}\,\psi^*_{\mathbf{k}}(\mathbf{r})J(\mathbf{r}-\mathbf{R}_i)\psi_{\mathbf{k}'}(\mathbf{r})\\
&= \mathbf{I}_i\cdot\boldsymbol{\sigma}\int d\mathbf{r}\,\psi^*_{\mathbf{k}}(\mathbf{r}+\mathbf{R}_i)J(\mathbf{r})\psi_{\mathbf{k}'}(\mathbf{r}+\mathbf{R}_i)\\
&= \mathbf{I}_i\cdot\boldsymbol{\sigma}e^{-i(\mathbf{k}-\mathbf{k}')\cdot\mathbf{R}_i}\int d\mathbf{r}\psi^*_{\mathbf{k}}(\mathbf{r})J(\mathbf{r})\psi_{\mathbf{k}'}(\mathbf{r})\\
&\equiv J(\mathbf{k},\mathbf{k}')\mathbf{I}_i\cdot\boldsymbol{\sigma}e^{-i(\mathbf{k}-\mathbf{k}')\cdot\mathbf{R}_i}
\end{aligned} \tag{6.32}$$

In deriving Eq.(6.32) we made use of the fact that since $\mathbf{R}_i$ is a lattice site,

[3] M. A. Ruderman and C. Kittel, *Phys. Rev.* **96**, 99 (1954).

the Bloch wavefunctions satisfy $\psi_{\mathbf{k}}(\mathbf{r}+\mathbf{R}_i) = e^{i\mathbf{k}\cdot\mathbf{R}_i}\psi_{\mathbf{k}}(\mathbf{r})$. If the exchange interaction has a very short range, $J(\mathbf{r}) = A\delta(\mathbf{r})$, then $J(\mathbf{k},\mathbf{k}') = A\psi^*_{\mathbf{k}}(0)\psi_{\mathbf{k}'}(0)$. A coupling between the impurity spins emerges at the second order of perturbation in J. Suppose the Fermi sea is filled up to a maximum wavevector $k_{\max}$. The second order correction to the energy of the system involving two different impurities at sites $\mathbf{R}_i$ and $\mathbf{R}_j$ is

$$\begin{aligned} H(\mathbf{R}_{ij}) &= \mathrm{Tr}_{\sigma}\int (d\mathbf{k}) n_{\mathbf{k}} \int (d\mathbf{k}') \frac{V^{(i)}(\mathbf{k},\mathbf{k}')V^{(j)}(\mathbf{k}',\mathbf{k})}{\epsilon^{(0)}_{\mathbf{k}} - \epsilon^{(0)}_{\mathbf{k}'}} \\ &= -\mathrm{Tr}_{\sigma}[(\mathbf{I}_i\cdot\boldsymbol{\sigma})(\mathbf{I}_j\cdot\boldsymbol{\sigma})]\int (d\mathbf{k}) n_{\mathbf{k}} \int (d\mathbf{k}') \frac{|J(\mathbf{k},\mathbf{k}')|^2}{\epsilon^{(0)}_{\mathbf{k}'} - \epsilon^{(0)}_{\mathbf{k}}} e^{-i(\mathbf{k}-\mathbf{k}')\cdot\mathbf{R}_{ij}}. \end{aligned} \tag{6.33}$$

Assume $J(\mathbf{k},\mathbf{k}')$ does not vary appreciably as a function of the wavevectors, and we replace it with a constant J. After this, we separately perform the angular integrations in $\mathbf{k}$ and $\mathbf{k}'$. Since the energy $\epsilon^{(0)}_{\mathbf{k}}$ depends only on the magnitude of $\mathbf{k}$, the angular integrations are quite simple. For example,

$$\int_0^{2\pi} d\phi \int_0^{\pi} d\theta \sin\theta e^{ik'R_{ij}\cos\theta} = 4\pi\frac{\sin(k'R_{ij})}{k'R_{ij}}.$$

We can also take the trace over the Pauli matrices now, and that gives $2(\mathbf{I}_i\cdot\mathbf{I}_j)$. At this stage we can write

$$\begin{aligned} H(\mathbf{R}_{ij}) &= -\frac{8J^2(\mathbf{I}_i\cdot\mathbf{I}_j)}{(2\pi)^4R_{ij}^2}\int_0^{k_{\max}} dk\, k\sin(kR_{ij}) \int_0^{\infty} dk' \frac{k'\sin(k'R_{ij})}{\epsilon^{(0)}_{\mathbf{k}'} - \epsilon^{(0)}_{\mathbf{k}}} \\ &= \frac{4m(\mathbf{I}_i\cdot\mathbf{I}_j)}{(2\pi)^4\hbar^2R_{ij}^2}\int_{-k_{\max}}^{k_{\max}} dk\, k\, e^{ikR_{ij}} \int_{-\infty}^{\infty} dk'\, k' \frac{e^{ik'R_{ij}}}{k'^2 - k^2}. \end{aligned}$$

Completing the momentum integrals we get

$$H(\mathbf{R}_{ij}) = \frac{2J^2m(\mathbf{I}_i\cdot\mathbf{I}_j)}{(2\pi)^3\hbar^2}\frac{2k_{\max}R_{ij}\cos(2k_{\max}R_{ij}) - \sin(2k_{\max}R_{ij})}{R_{ij}^4}. \tag{6.34}$$

6.2.2.1 Giant magnetoresistance in multilayers

Perhaps the most striking manifestation of the RKKY interaction is the giant magnetoresistance of ferromagnetic metal - nonmagnetic metal multilayers. These multilayer devices are made of alternating sheets of ferromagnetic metal (e.g. Fe, Co, Ni) and nonmagnetic metal (e.g. Cu, Cr, Au). When the thickness of the nonmagnetic sheets is increased, the exchange coupling between the ferromagnetic sheets is observed to alternate between ferromagnetic and antiferromagnetic. This suggests an RKKY mechanism, where the spins in a given ferromagnetic layer induce an oscillating magnetization in the adjoining nonmagnetic layer which is conveyed to the next ferromagnetic layer.

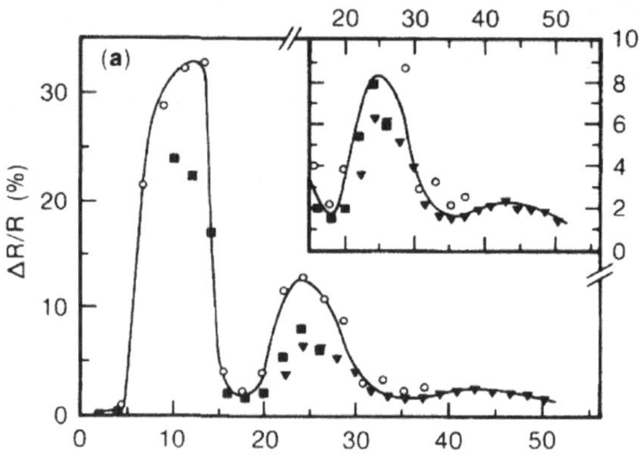

FIGURE 6.4: Giant saturation magnetoresistance of a Fe-Cr multilayer structure as a function of the thickness (in angstrom) of the nonmagnetic Cr layer. The resistance is maximum when the neighboring Fe layers are antiferromagnetically aligned and minimum when ferromagnetically aligned. Source: S. S. P. Parkin, N. More and K. P. Roche, *Phys. Rev. Lett.* **64**, 2304 (1990).

However the observed period of oscillation of the exchange coupling is typically many times larger than π/k_F. It turns out that the mechanism is still of the RKKY-type indirect exchange but not in the naive way Eq.(6.34) would suggest.[4] To calculate the RKKY coupling of a particular spin on one of the ferromagnetic layers with the spins on a neighboring ferromagnetic layer, one needs to sum over contributions from the spins in that neighboring layer. It is important not to approximate the atoms in the neighboring layer as a continuum, for the interatomic spacing in the ferromagnetic layers is not small compared to the Fermi wavelength. Upon performing this discrete sum, one finds that the period of RKKY oscillations is decreased by a factor of $\sqrt{1-(G/2k_F)^2}$, where G is a two-dimensional reciprocal lattice vector in the ferromagnetic layer.

In the actual devices, the thickness of the nonmagnetic metal is adjusted so that neighboring ferromagnetic sheets couple antiferromagnetically. If a current is passed in a direction normal to the sheets, the alternating polarization of the ferromagnetic sheets offers a large resistance. Now if a strong enough magnetic field is applied to the multilayer so as to align the magnetic layers ferromagnetically, the resistance becomes much lower. The change in resistance can be as much as about 50% in fields of the order of a tesla; hence the term giant magnetoresistance[5] (see Figure 6.4).

[4] P. Bruno and C. Chappert, *Phys. Rev. B* **46**, 261 (1992).
[5] See S. S. P. Parkin, N. More and K. P. Roche, *Phys. Rev. Lett.* **64**, 2304 (1990).

6.3 Hartree and Hartree-Fock approximations

We have learned how the conduction electrons in a metal screen the Coulomb interaction. One is naturally interested in the question as to how the noninteracting electron energies $\epsilon_{\mathbf{k}}^{(0)}$ in turn are affected by the introduction of the Coulomb interaction. We discuss two approximations commonly used in this context. These are, respectively, the Hartree and the Hartree-Fock approximations.

6.3.1 Hartree approximation

Consider the Schrödinger equation for an $N-$electron system in the metal,

$$\sum_{i=1}^{N}\left(\frac{\mathbf{p}_i^2}{2m}-\frac{Ze^2}{4\pi\epsilon_0}\sum_{\mathbf{R}_j}\frac{1}{|\mathbf{r}_i-\mathbf{R}_j|}\right)\Psi+\frac{e^2}{8\pi\epsilon_0}\sum_{i\neq j}\frac{1}{|\mathbf{r}_i-\mathbf{r}_j|}\Psi=E\Psi. \tag{6.35}$$

The first term on the left hand side is the kinetic energy of the electrons, the second term is the electron-ion interaction, and the last term is the electron-electron interaction. The total wavefunction Ψ is a function of the N coordinates and spins, and, in general, is not equal to a product of single-particle wavefunctions. Directly solving for Ψ when N is large is a practically impossible task.

The Hartree approximation attempts to simplify the problem to that of a single electron moving in an effective potential of all the other charges. The first approximation is that the electron number density $\rho(\mathbf{r})$ can be regarded as a smoothly varying quantity. That enables us to write the mutual interaction energy of the electrons (i.e., the last term on the left hand side of Eq.(6.35)) as

$$U_{ee}(\mathbf{r})=-\frac{e}{4\pi\epsilon_0}\int d\mathbf{r}'\rho(\mathbf{r}')\frac{1}{|\mathbf{r}-\mathbf{r}'|}. \tag{6.36}$$

The second approximation is that the electron wave functions are assumed to be independent. Thus instead of the many-particle wavefunction Ψ, we work with single-particle wavefunctions ψ_i. In this approximation, the charge density corresponding to ψ_i is $\rho_i(\mathbf{r})=-e|\psi_i(\mathbf{r})|^2$. The total electron charge density is just the sum, $\rho(\mathbf{r})=\sum_i\rho_i(\mathbf{r})$. The Hartree equation for the ith state is

$$-\frac{\hbar^2}{2m}\nabla^2\psi_i(\mathbf{r})-\frac{Ze^2}{4\pi\epsilon_0}\sum_{\mathbf{R}_j}\frac{1}{|\mathbf{r}-\mathbf{R}_j|}\psi_i(\mathbf{r})+$$

$$+\left(\frac{e^2}{4\pi\epsilon_0}\sum_j\int d\mathbf{r}'|\psi_j(\mathbf{r}')|^2\frac{1}{|\mathbf{r}-\mathbf{r}'|}\right)\psi_i(\mathbf{r})=E_i\psi_i(\mathbf{r}). \tag{6.37}$$

These are a set of N non-linear differential equations that are usually solved by numerical iteration. The total wavefunction Ψ in the Hartree approximation is simply the product of the single-particle wavefunctions,

$$\Psi = \psi_1(\mathbf{r}_1, \sigma_1)\psi_2(\mathbf{r}_2, \sigma_2)\ldots\psi_N(\mathbf{r}_N, \sigma_N). \tag{6.38}$$

If we minimize the expectation value of the Hamiltonian, $\langle H\rangle = \int d\mathbf{r}\,\Psi^* H\Psi$, subject to the constraint that the wavefunction is normalized, i.e., $\int d\mathbf{r}\,\Psi^*\Psi = 1$, we will obtain the Hartree equation, Eq.(6.37).

The Hartree approach, being an independent electron treatment, misses out on physical effects that depend on the fact that the electrons are fermions obeying the Pauli principle.

6.3.2 Hartree-Fock approximation

The Hartree-Fock approach also attempts to simplify the many-body problem, and unlike the earlier Hartree approach, builds in the Pauli principle. In the Hartree-Fock approximation, the total wavefunction is taken as the anti-symmetrized product of the single-particle wavefunctions,

$$\Psi(\mathbf{r}_1, \sigma_1; \mathbf{r}_2, \sigma_2; \cdots; \mathbf{r}_N, \sigma_N) = \mathcal{N}\det\begin{pmatrix} \psi_1(\mathbf{r}_1,\sigma_1) & \psi_1(\mathbf{r}_2,\sigma_2) & \cdot\;\cdot & \psi_1(\mathbf{r}_N,\sigma_N) \\ \psi_2(\mathbf{r}_1,\sigma_1) & \psi_2(\mathbf{r}_2,\sigma_2) & \cdot\;\cdot & \psi_2(\mathbf{r}_N,\sigma_N) \\ \cdot & \cdot & \cdot\;\cdot & \cdot \\ \cdot & \cdot & \cdot\;\cdot & \cdot \\ \psi_N(\mathbf{r}_1,\sigma_1) & \psi_N(\mathbf{r}_2,\sigma_2) & \cdot\;\cdot & \psi_N(\mathbf{r}_N,\sigma_N) \end{pmatrix}. \tag{6.39}$$

$\mathcal{N}$ is the normalization. Note that this is an approximation because it involves our assuming that the $N-$particle wavefunction can be factorized into a product of single-particle wavefunctions. If we use Ψ as shown in Eq.(6.39) as the trial wavefunction and minimize $\int d\mathbf{r}\,\Psi^* H\Psi$ subject to the constraint $\int d\mathbf{r}\,\Psi^*\Psi = 1$, we will obtain the Hartree-Fock equations,

$$\begin{aligned} &-\frac{\hbar^2}{2m}\nabla^2\psi_i(\mathbf{r}) - \frac{Ze^2}{4\pi\epsilon_0}\sum_{\mathbf{R}_j}\frac{1}{|\mathbf{r}-\mathbf{R}_j|}\psi_i(\mathbf{r}) \\ &+\left(\frac{e^2}{4\pi\epsilon_0}\sum_j\int d\mathbf{r}'|\psi_j(\mathbf{r}')|^2\frac{1}{|\mathbf{r}-\mathbf{r}'|}\right)\psi_i(\mathbf{r}) \\ &-\frac{e^2}{4\pi\epsilon_0}\sum_j\delta_{\sigma_i,\sigma_j}\int d\mathbf{r}'\frac{1}{|\mathbf{r}-\mathbf{r}'|}\psi_j^*(\mathbf{r}')\psi_i(\mathbf{r}')\psi_j(\mathbf{r}) = E_i\psi_i(\mathbf{r}). \end{aligned} \tag{6.40}$$

The last term on the left hand side of Eq.(6.40) is known as the *exchange* term. Exchange is what distinguishes Hartree-Fock from Hartree, and it is a direct consequence of requiring the many-body wavefunction to obey the

Pauli principle. The Hartree-Fock equations are in general very difficult to solve because the form of the exchange term is not local, $V(\mathbf{r})\psi(\mathbf{r})$. Instead it has a has a *nonlocal* form, $\int U(\mathbf{r}-\mathbf{r}')\psi(\mathbf{r}')$.

In a real metal, the electrons move in the potential of the positively charged ions, and solving the Hartree-Fock equations in the periodic potential of the ions is very complicated. However in case of a free electron gas, the Hartree-Fock equations can be solved. And using the free electron gas solution as a starting point, one can say a few things about the electron gas in a lattice. Therefore we will discuss briefly the Hartree-Fock treatment for free electrons. We can choose plane waves as trial single-particle wavefunctions for the free electron gas:

$$\psi_i(\mathbf{r}) = \frac{1}{\sqrt{V}} e^{i\mathbf{k}_i \cdot \mathbf{r}}. \tag{6.41}$$

A uniform electronic density is associated with the plane waves. In the free electron gas, the effect of the ions is modeled by a uniform positive charge density. It is easy to see that in this "uniform" approximation, the electron-electron interaction energy and the electron-ion interaction energy cancel each other out. The only contribution to the electron potential energy then comes from the exchange part of the interaction. We need the Fourier transform of the Coulomb interaction in the exchange term:

$$\frac{e^2}{4\pi\epsilon_0} \frac{1}{|\mathbf{r}-\mathbf{r}'|} = \frac{e^2}{\epsilon_0} \int (d\mathbf{q}) \frac{1}{q^2} e^{i\mathbf{q}\cdot(\mathbf{r}-\mathbf{r}')}. \tag{6.42}$$

When we make use of Eq.(6.42) in the Hartree-Fock equation, we find that the plane waves given by Eq.(6.41) are a solution, and the energy $E_{\mathbf{k}}$ corresponding to wavevector $\mathbf{k}$ is given by

$$\begin{aligned} E_{\mathbf{k}} = \epsilon_{\mathbf{k}}^{(0)} + E_{\mathbf{k}}^{\text{exch}} &= \frac{\hbar^2 k^2}{2m} - \frac{e^2}{\epsilon_0} \int_{k'<k_F} (d\mathbf{k}') \frac{1}{|\mathbf{k}-\mathbf{k}'|^2} \\ &= \frac{\hbar^2 k^2}{2m} - \frac{e^2 k_F}{2\pi^2 \epsilon_0} F(k) = \frac{\hbar^2 k^2}{2m} - \frac{\hbar^2}{2m} \left(\frac{4k_F}{\pi a_B} \right) F(k), \end{aligned} \tag{6.43}$$

where

$$F(k) = \frac{1}{2} + \frac{1-(k/k_F)^2}{4(k/k_F)} \ln \left| \frac{1+(k/k_F)}{1-(k/k_F)} \right|. \tag{6.44}$$

This result is for zero temperature of course. Note that the integral over the wavevector k' runs up to a maximum value k_F because a given electron interacts with the rest of the electrons that can have a maximum wavevector k_F. Also, the function F that appears in the exchange energy is familiar to us. Eq.(6.19) shows that the Lindhard polarizability $\Pi(\mathbf{q})$ in three dimensions is proportional to $F(q/2)$.

Figure 6.5 shows plots of the free electron energy and the electron energy

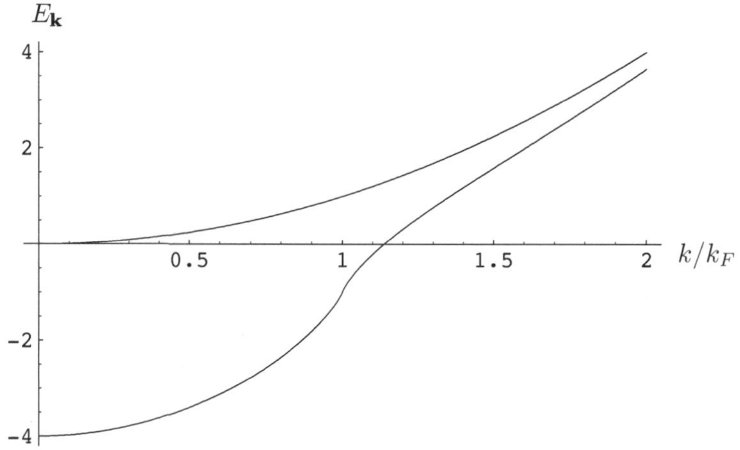

FIGURE 6.5: Plots of the electron energy as a function of the wavevector k for a free, noninteracting electron gas the energy, and for the interacting electron gas the Hartree-Fock approximation. The lower curve shows the Hartree-Fock value. The energy is measured in units of the Fermi energy of noninteracting electrons, $\hbar^2 k_F^2/2m$, and the wavevector is measured in units of k_F. The plot is for a specific value of electron density, $k_F a_B = 4/\pi$. Note the kink in the Hartree-Fock dispersion at $k = k_F$. The slope $dE_\mathbf{k}/d\mathbf{k}$ is vertical at $k = k_F$ which makes the Fermi velocity infinite. The bare Coulomb interaction was used and screening effects were ignored in this plot.

$E_{\mathbf{k}}$ in the Hartree-Fock approximation as a function of the wavevector k. In the figure, one can make out a kink in the Hartree-Fock dispersion at $k = k_F$. By differentiating Eq.(6.43) with respect to $\mathbf{k}$, we find that the Fermi velocity, which is proportional to $dE_{\mathbf{k}}/d\mathbf{k}$, has a logarithmic singularity at $k = k_F$. This singularity at the Fermi surface is a consequence of the long-range $(1/r)$ nature of the Coulomb interaction.

However in a metal, we should take into account screening effects in the Coulomb interaction of the electrons. One rough prescription is to estimate the exchange energy in Hartree-Fock using the screened Coulomb interaction instead of the bare Coulomb interaction. A consequence of this modification is that we will no longer get a singular Fermi velocity. If one uses the Thomas-Fermi screened Coulomb interaction,

$$\frac{e^2}{\epsilon_0}\frac{1}{\mathbf{k}^2 + (1/\lambda_{TF})^2},$$

one finds that the Hartree-Fock exchange energy,

$$E_{\mathbf{k}}^{TF} = \epsilon_{\mathbf{k}}^{(0)} - \frac{e^2}{\epsilon_0}\int_{k'<k_F} (d\mathbf{k}')\frac{1}{(\mathbf{k}-\mathbf{k}')^2 + (1/\lambda_{TF})^2}, \tag{6.45}$$

is not very significant compared to the noninteracting value.

If however we also consider the effect of the motion of the positively charged ions on the dielectric constant, and therefore, on the exchange energy, it turns out that the exchange term makes an interesting difference to the electron energies near the Fermi surface. How that happens will be discussed in the following section.

6.3.2.1 Fermi liquid parameters

In Eq.(6.43), we can read off a Fermi liquid parameter $f(\mathbf{k},\sigma;\mathbf{k}',\sigma')$ from the exchange term:

$$f(\mathbf{k},\sigma;\mathbf{k}',\sigma') = -\delta_{\sigma,\sigma'}\frac{e^2}{\epsilon_0|\mathbf{k}-\mathbf{k}'|^2}, \tag{6.46}$$

which implies

$$f^s(\mathbf{k},\mathbf{k}') = f^a(\mathbf{k},\mathbf{k}') = -\frac{e^2}{2\epsilon_0|\mathbf{k}-\mathbf{k}'|^2}. \tag{6.47}$$

These values correspond to the bare Coulomb interaction. The RHS of Eq. (6.46) is proportional to the well-known Rutherford scattering amplitude for the Coulomb problem. This suggests a connection between the Landau f-functions and the scattering amplitudes. We will see in Ch. 9 that f is closely related to the scattering amplitude (although it is not directly proportional to f.)

6.3.2.2 Compressibility and susceptibility

We can obtain the compressibility κ of the electron gas using the relation $1/\kappa = n^2(\partial\mu/\partial n)_{V,T}$, where $n = k_F^3/3\pi^2$ is the electron density. Considering a bare Coulomb interaction, the chemical potential μ^{HF} is simply the Hartree-Fock energy of the electron, shown in Eq.(6.43), evaluated for $k = k_F$:

$$\mu^{HF} = \frac{\hbar^2 k_F^2}{2m} - \frac{e^2 k_F}{4\pi^2 \epsilon_0}. \tag{6.48}$$

The second term in Eq.(6.48) is purely due to interaction, and we can recover the noninteracting value by taking the limit $e \to 0$. We can show that the Hartree-Fock compressibility κ^{HF} as measured the noninteracting value κ is given by

$$\begin{aligned} \frac{\kappa^{HF}}{\kappa} &= \frac{(\partial\mu/\partial n)_{V,T}}{(\partial\mu^{HF}/\partial n)_{V,T}} = \frac{\frac{\hbar^2 k_F}{m}}{\left(\frac{\hbar^2 k_F}{m} - \frac{e^2}{4\pi^2\epsilon_0}\right)} \\ &= \frac{1}{1 - \frac{me^2}{4\pi^2\epsilon_0\hbar^2 k_F}} = \frac{1}{1 - \frac{(r_{ee}/a_B)}{6.02}}, \end{aligned} \tag{6.49}$$

where r_{ee} is the electron-electron separation. Thus a three-dimensional electron gas in the Hartree-Fock approximation (and a bare Coulomb interaction) is unstable against density fluctuations when the electron density is low enough such that $r_{ee}/a_B \geq 6.02$. This shows that a homogeneous electron gas at low enough densities, and moving in a uniform positive background, will spontaneously phase-separate.

We will get the same result for the magnetic susceptibility because $\kappa^{HF}/\kappa = (1 + F_1^s/3)/(1 + F_0^s)$, and $\chi^{HF}/\chi = (1 + F_1^s/3)/(1 + F_0^a)$, and Eq.(6.47) tells us that $f^s = f^a$. Hence

$$\frac{\chi^{HF}}{\chi} = \frac{1}{1 - \frac{me^2}{4\pi^2\epsilon_0\hbar^2 k_F}} = \frac{1}{1 - \frac{(r_{ee}/a_B)}{6.02}}. \tag{6.50}$$

The electron gas in the Hartree-Fock approximation is thus susceptible to a ferromagnetic transition at low densities.

6.4 Electrons in metal crystals

Many of the properties of an electron gas in a metal can be very different from a free electron gas because of additional effects of the periodic potential provided by the lattice of ions.

6.4.1 Electrons in a rigid lattice

Let us first discuss the case where the ions stay rigidly fixed to their lattice positions. Many important effects of the presence of the lattice already appear at this stage. The wavefunction of an electron in a periodic potential has the form

$$\psi_{s,\mathbf{k}}(\mathbf{r}) = e^{i\mathbf{k}\cdot\mathbf{r}} u_{s,\mathbf{k}}(\mathbf{r}). \tag{6.51}$$

Eq.(6.51) is called a Bloch function. $u_{s,\mathbf{k}}(\mathbf{r})$ is invariant under translation by a lattice vector,

$$u_{s,\mathbf{k}}(\mathbf{r}+\mathbf{a}) = u_{s,\mathbf{k}}(\mathbf{r}); \tag{6.52}$$

such a form ensures that the probability density $|\psi_{s,\mathbf{k}}|^2$ is invariant under lattice translations.[6] The label $\mathbf{k}$ denotes not the actual momentum of the electron but a new quantity called the crystal momentum. We can directly check, using Heisenberg's equations of motion, that if a particle moves in a potential that is not uniform, the momentum of the electrons is not conserved. The crystal momentum in many ways resembles the actual momentum. Suppose we consider two electrons. If we translate the two electrons by the same lattice vector, the wavefunction will, as before, change by an amount $e^{i\mathbf{k}\cdot\mathbf{a}}$. If the two electrons are very far from each other, then the total wavefunction can be written as a product of the individual wavefunctions with individual crystal momenta $\mathbf{k}_1$ and $\mathbf{k}_2$ respectively. The change in the "product" wavefunction upon the same translation is $e^{i\mathbf{k}_1\cdot\mathbf{a}}e^{i\mathbf{k}_2\cdot\mathbf{a}}$. Then we can say

$$\mathbf{k} = \mathbf{k}_1 + \mathbf{k}_2 + \mathbf{G},$$

where $\mathbf{G}$ is a reciprocal lattice vector, $e^{i\mathbf{G}\cdot\mathbf{a}} = 1$. From here, it is easy to see that in a collision of two electrons in a periodic field, the total crystal momentum is conserved to within a reciprocal lattice vector,

$$\mathbf{k}_1 + \mathbf{k}_2 = \mathbf{k}_1' + \mathbf{k}_2' + \mathbf{G}. \tag{6.53}$$

It is possible to have infinitely many different discrete values of energy $\epsilon_{s,\mathbf{k}}$ corresponding to the same crystal momentum $\mathbf{k}$. These states have been labelled above by the indices n which denote different "bands". $\epsilon_{s,\mathbf{k}}$ is referred to as the energy of the sth band. The range of values of energy in each band is finite. Different bands may either have no common range of energies over which they intersect, or they may overlap in part. In the former case, the two bands are separated by a "band gap". When a part of two bands have a common

---

[6] Actually, the invariance of $|\psi_{s,\mathbf{k}}|^2$ under lattice translations would require the form $e^{i\varphi(\mathbf{r})}u_{s,\mathbf{k}}(\mathbf{r})$. In order to get the appropriate form $\varphi(\mathbf{r}) = \mathbf{k}\cdot\mathbf{r}$, we need to make use of the additional fact that two successive lattice translations, made in any order, would lead to the same wavefunction as one where one makes a single translation, equal to the sum of the above two translations. See for example Ashcroft and Mermin, *Solid State Physics.*

energy, they may or may not correspond to the same crystal momentum. If at the intersection the two bands have the same wavevector, the bands will in general mix.

The energy of any band is invariant under translations by a reciprocal lattice vector $\mathbf{G}$,

$$\epsilon_{s,\mathbf{k}} = \epsilon_{s,\mathbf{k}+\mathbf{G}}. \tag{6.54}$$

One can show that the mean value of the electron velocity has the same form as for free electrons moving in a uniform potential:

$$\mathbf{v}_s = \frac{1}{\hbar}\frac{\partial \epsilon_{s,\mathbf{k}}}{\partial \mathbf{k}}. \tag{6.55}$$

In a crystal, the Fermi surface can be quite complicated. If we regard the lattice as a perturbation of the free electron gas in a uniform positive background, the ground state of the electrons in the lattice will correspond to some excited state of the electron gas in the absence of the lattice. The new ground state will reflect the symmetry of the lattice (perturbation). Clearly, the Fermi surface need not be spherical. Even though the Fermi surface will still be invariant with respect to simultaneous reflection about the origin and reversal of spins (time reversal symmetry), it is possible to have non-trivial effects of the lattice. For example, in crystals that have no center of symmetry, the Fermi surfaces corresponding to the two spin species are different. In general, the Fermi surface will be composed of parts of many different bands. It might have many sheets that can have an open or closed geometry, and the sheets may overlap.

At $T = 0$, the particles of a noninteracting Fermi gas in a lattice will occupy the lowest energy states (labelled by the band index s, the crystal momentum $\mathbf{k}$, and possibly the spin) and fill up to a maximum energy $\mu = \epsilon_F$. The closed surfaces can either enclose filled states ($\epsilon < \epsilon_F$) or empty states ($\epsilon > \epsilon_F$). In the former case, an excitation will take particles from inside the volume to outside. This is called an *electron-type* Fermi surface. In the latter case, we can imagine the volume to be filled with holes, and an excitation will take holes from within the volume to its exterior. Such a Fermi surface is called *hole-type.* This is convenient because when the Fermi surface encloses empty states, the enclosed volume is usually much smaller than the volume outside. Figure 6.6 shows the Fermi surfaces of potassium and aluminium.

For the homogeneous Fermi liquid, we argued in Sec. 1.2 that the Fermi momentum k_F is unaffected by the mutual interaction of the fermions, or in other words, the volume of the Fermi sphere in momentum space is conserved: $(4\pi k_F^3/3) = 4\pi^3 n$. A similar result referred to as Luttinger's theorem exists for electrons in a crystalline metal. We need to define what we mean by the volume enclosed by a Fermi surface in a lattice. Consider a unit cell in the reciprocal space. Let τ_F be the volume in reciprocal space enclosed by the Fermi surface on the side of the occupied states, that is, for $\epsilon < \epsilon_F$. If there are two different overlapping sheets (bands) of the Fermi surface, then the

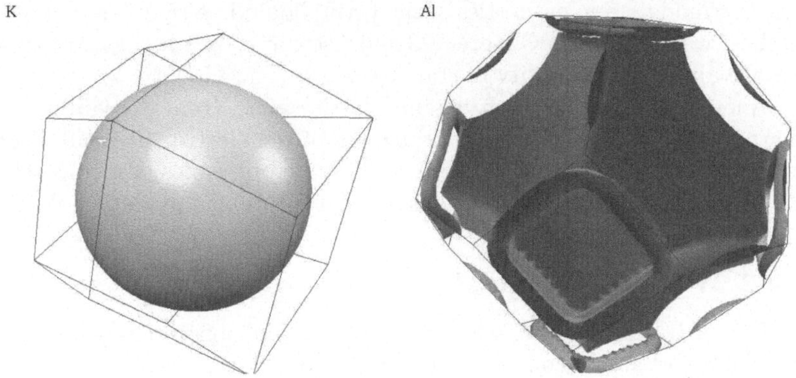

FIGURE 6.6: The Fermi surface of potassium is nearly spherical and consists of just one band while that of aluminium is more complicated and consists of two bands. Source: Fermi Surface Database, University of Florida, `http://www.phys.ufl.edu/fermisurface/`

volumes enclosed by them should be reckoned independently, and added. In principle, τ_F can exceed the volume of the unit cell in the reciprocal lattice. Denote the number of electrons per unit lattice cell as N. This includes both conduction and core electrons. Suppose a number l of the bands is completely filled up. The complete occupation of l bands contributes $2l$ electrons to the unit cell of the reciprocal lattice, and the remaining $N - 2l$ is the number of conduction electrons per unit cell of the reciprocal lattice. Suppose V^{cell} is the volume of the unit lattice cell in real space. The Luttinger theorem states that

$$2\tau_F = (N - 2l)/V^{\text{cell}}. \tag{6.56}$$

The main point of the Luttinger theorem, as we shall see later in Sec. 9.4, is that it holds even for interacting electrons.[7] The low energy excitations of the Fermi liquid in a lattice are particle-holes pairs near the Fermi surface. The density of states at the Fermi surface needs to be defined somewhat differently because the Fermi surface is no longer a sphere. Let dS be an element of area in $\mathbf{k}-$space at the Fermi surface. The contribution of this element to the density of states is $(1/4\pi^3)dS/\hbar|\mathbf{v}_F(\mathbf{k})| = (1/4\pi^3)dS/\hbar v_F(\mathbf{k})$, where $\mathbf{v}_F(\mathbf{k}) = (1/\hbar)\partial E_{\mathbf{k}}/\partial \mathbf{k}$ is a vector normal to the Fermi surface whose magnitude is the Fermi velocity at that point of the surface; $E_{\mathbf{k}}$ is the energy including Coulomb and other electron-electron interactions. The total density of states $\nu(0)$ is

$$\nu(0) = \frac{1}{4\pi^3} \int_{\epsilon_{\mathbf{k}}=\epsilon_F} \frac{dS}{\hbar v_F(\mathbf{k})}. \tag{6.57}$$

[7] J. M. Luttinger, *Phys. Rev.* **119**, 1153 (1960).

The thermodynamic properties of the Fermi liquid in a rigid lattice are similar to those we obtained in Chapter 1.3 and elsewhere, except that now we should use the appropriate density of states given by Eq.(6.57).

Another class of new features in a lattice arises from the vibration of the charged ions, which affects the properties of the electron gas and vice versa. In the rest of this section, we shall study some properties of metals related to the interaction of lattice vibrations, or phonons, and conduction electrons.

6.4.2 Phonons in metals

Let us begin our discussion with phonons in non-ionic crystals modeled by a lattice of atoms connected to their nearest neighbors by springs. Solving the equations of motion for the atoms gives us the normal modes. For a one-dimensional chain of atoms of mass M and separation a connected to their nearest neighbors with spring constant K, the phonon frequency corresponding to wavevector $\mathbf{k}$ is given by

$$\omega_{\mathbf{k}}^2 = 2(K/M)(1 - \cos ka). \tag{6.58}$$

Thus at small wavevectors $ka \ll 1$, the phonon frequency is proportional to the wavevector.

In a metal, the atoms constituting the lattice become positively charged after losing conduction electrons. The lattice vibrations are then influenced by the long-range Coulomb interaction of the ions with each other, and by the screening motion of the conduction electrons. In the absence of screening effects of the conduction electrons, the positively charged ions can undergo density oscillations at the ionic plasma frequency Ω_p given by

$$\Omega_p^2 = \frac{n_i(Ze)^2}{M\epsilon_0}, \tag{6.59}$$

where n_i is the density of ions, Z the ionic charge, and M the ionic mass. Note that these oscillations arise from long-range Coulomb interactions of the ions, and do not depend on the restoring forces provided by interatomic potential of the neutral atoms. We can obtain Eq.(6.59) by considering the electrostatic restoring force on a sheet of ions in the lattice displaced by an amount $\mathbf{x}$ from its equilibrium position:

$$M\ddot{\mathbf{x}} = Ze\mathbf{E} = -\frac{n_i(Ze)^2}{\epsilon_0}\mathbf{x}. \tag{6.60}$$

Here we immediately run into a contradiction with the experimental reality since one can observe directly in neutron scattering experiments, and indirectly through T^3 temperature dependence of the specific heat, that the frequency of lattice vibrations in metals vanishes linearly with wavevector k.

The problem is that we have not yet considered the effect of the conduction electrons. We know that in a medium the macroscopic electric fields are reduced from their bare values by a factor equal to the dielectric function. Therefore in the equation of motion, Eq.(6.60), we should modify the electric field appropriately. For a lattice vibration with wavevector $\mathbf{k}$, the correct restoring electric field should be $\mathbf{E}/\varepsilon(\mathbf{k})$, where $\mathbf{E}$ is the bare field in the absence of electrons. Consequently, the ionic plasma frequency in Eq.(6.59) should be modified to

$$\omega(\mathbf{k})^2 = \frac{\Omega_p^2}{\varepsilon(\mathbf{k})}. \tag{6.61}$$

As $\mathbf{k} \to 0$, the dielectric function should approach the Thomas-Fermi value:

$$\varepsilon(\mathbf{k}) \approx 1 + \frac{e^2\nu(0)}{\epsilon_0 \mathbf{k}^2} = 1 + \frac{1}{(\lambda_{TF}\mathbf{k})^2}. \tag{6.62}$$

Using $\Omega_p^2 = n_i(Ze)^2/M\epsilon_0$ and $\nu(0) = mk_F/\pi^2\hbar^2$ (in 3D), we find that for small wavevectors such that $k\lambda_{TF} \ll 1$, the ionic plasma frequencies are

$$\omega(\mathbf{k})^2 = \Omega_p^2 \lambda_{TF}^2 k^2 = \frac{n_i Z^2}{M}\frac{\pi^2\hbar^2}{mk_F}k^2.$$

Thus the dispersion relation is linear as desired in the $\mathbf{k} \to 0$ limit. Since each lattice atom contributes Z conduction electrons, the conduction electron density n and ionic density n_i are related through $n = Zn_i$. Making use of $n_i = n/Z = k_F^3/3Z\pi^2$, we arrive at the following relation for the (long wavelength) dispersion relation for the vibration of the ions interacting through the screened Coulomb potential:

$$\omega(\mathbf{k})^2 = \frac{Zp_F^2}{3Mm}k^2 \equiv \frac{Zmv_F^2}{3M}k^2. \tag{6.63}$$

Eq.(6.63) is known as the Bohm-Staver[8] relation. Since the electron mass is typically 10^{-4} to 10^{-5} times smaller than the ionic mass, the speed of the Bohm-Staver phonons is about a hundred times smaller than v_F. This speed roughly agrees with experiments.

6.4.2.1 Kohn anomalies

Electrostatic screening by electrons affects the vibration of the ions in the metal even at large wavevectors. However, at wavevectors which are not small compared to $1/\lambda_{TF}$, we can no longer use the Thomas-Fermi dielectric function in Eq.(6.61) for obtaining the dispersion relation for the phonons. A more accurate treatment should use the Lindhard dielectric function. It was pointed out by Kohn[9] that since the Lindhard dielectric function is singular at wavevectors $k = 2k_F$, this singularity should be reflected in the phonon

[8] D. Bohm and T. Staver, *Phys. Rev.* **84**, 836 (1950).
[9] W. Kohn, *Phys. Rev. Lett.* **2**, 393 (1959).

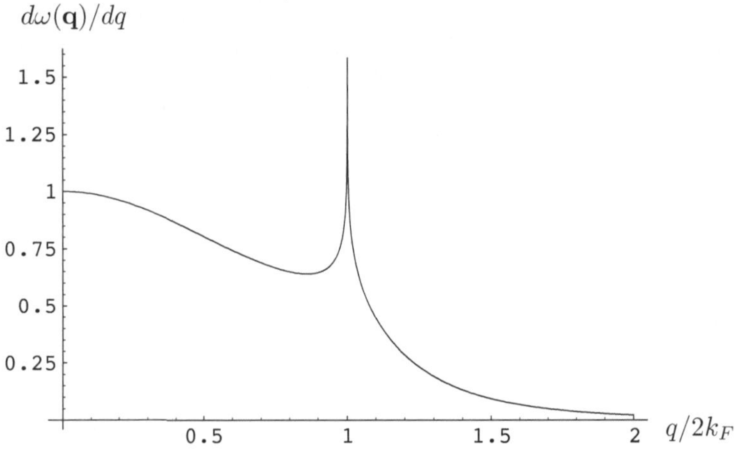

FIGURE 6.7: Illustration of the Kohn anomaly in three dimensions. The plot is of the slope, $d\omega/dq$, of the phonon dispersion relation as given by Eq.(6.61) using the Lindhard value for the dielectric function $\varepsilon(\mathbf{q})$. The slope is normalized against the (Bohm-Staver) value at $q = 0$. The singular behavior at $q = 2k_F$ is the Kohn anomaly.

spectrum as well. The result is weak singularities in $d\omega(\mathbf{k})/d\mathbf{k}$ for values of k that correspond to extremal diameters of the Fermi surface.

These weak singularities in the phonon spectrum can be used to obtain the shape of the Fermi surface. Highly accurate neutron scattering measurements[10] are required to get the phonon dispersion relation and reveal the weak singularities. Figure 6.7 shows a plot of $d\omega/dq$ as a function of q, where we have assumed for simplicity the free electron dispersion$\epsilon_{\mathbf{k}}^{(0)} = \hbar^2 k^2/2m$ in the calculation of the Lindhard polarizability. This is a reasonable approximation if k_F lies well within the first Brillouin zone. Otherwise we should use more realistic electron dispersions that are consistent with the periodicity of the lattice. An example would be $\epsilon_{\mathbf{k}}^{(0)} = t(3 - \sum_{\mathbf{a}} \cos(\mathbf{k} \cdot \mathbf{a}))$ for a three dimensional cubic lattice. The singular behavior at $q = 2k_F$ is known as the Kohn anomaly. If k_F lies outside the first Brillouin zone, then the Kohn anomaly will not exist in the acoustic phonon branch, but it might exist in the optic phonon branch, if any.

In one dimension, the Kohn anomaly is closely connected with a phenomenon known as the Peierls instability. Peierls showed that a one dimensional lattice with one conduction electron per lattice atom is unstable against

[10]R. Stedman *et al.*, *Phys. Rev.* **162**, 545 (1967).

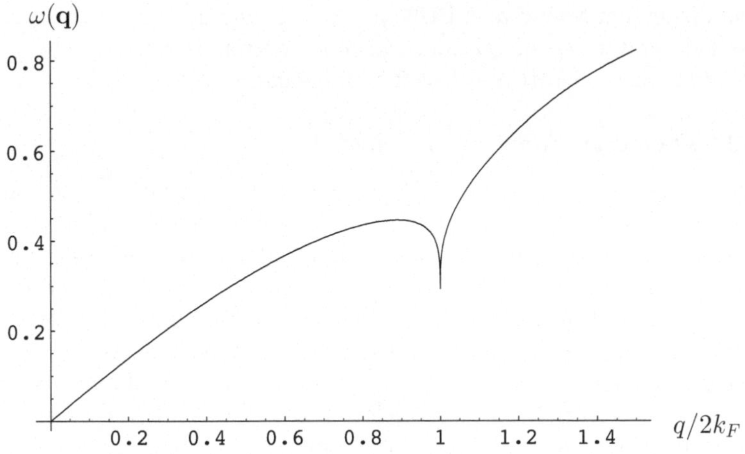

FIGURE 6.8: Plot of the phonon dispersion $\omega(\mathbf{q})$ in one dimension as given by Eq.(6.65). The Kohn anomaly in one dimension has the opposite sign compared to the three dimensional case. At $q = 2k_F$, the phonon frequency vanishes logarithmically; this is related to the Peierls instability.

a spontaneous distortion where neighboring atoms get displaced in opposite directions. This distortion corresponds to a wavevector $q = 2k_F = \pi/a$.[11]

Now consider the Kohn anomaly in one dimension. The Lindhard polarizability in one dimension is

$$\Pi(\mathbf{q}) = -\frac{e^2\nu(0)k_F}{q}\ln\left|\frac{1+q/2k_F}{1-q/2k_F}\right|. \qquad (6.64)$$

The polarizability diverges as $q \to 2k_F$. As a result, the phonon frequencies given by

$$\omega(\mathbf{q})^2 = \frac{\Omega_p^2}{\varepsilon(\mathbf{q})} = \frac{\Omega_p^2 q^2}{q^2 + (1/\lambda_{TF})^2(k_F/q)\ln\left|\frac{1+q/2k_F}{1-q/2k_F}\right|} \qquad (6.65)$$

become vanishingly small as $q \to 2k_F$. In other words, if we modulate the position of the lattice ions with a periodic wave with wavevector $q = 2k_F$, the modulation will remain "frozen" and not change with time. Figure 6.8 shows a plot of the phonon dispersion in one dimension.

This static periodic distortion where neighboring atoms have the opposite displacement leads to a doubling of the size of the unit cell from the earlier a to the new value $2a$. It also leads to the appearance of band gaps in the

[11]The Fermi wavevector $k_F = \pi/2a$ because each crystal momentum eigenstate can accommodate two electrons corresponding to the two spin species.

electron dispersion at the *new* Brillouin zone boundaries $k = \pm\pi/2a$.[12] Thus a half-filled band in one dimension, which one normally expects to be a metal, can spontaneously distort and become an insulator.

6.4.2.2 Dielectric function of a metal

When a metal is subjected to a small external potential $\phi^{(0)}$, not only the electrons but also the ions will rearrange themselves in an attempt to screen the external potential. The external and induced charge densities then create the total potential ϕ. Of course the ions are unable to move around freely to the same extent as the electrons do because the ions are much heavier, and the lattice is held together by the interatomic forces. Nevertheless, the ions may undergo small displacements about their equilibrium positions when subjected to an external potential. So in the relation $\phi^{(0)}(\mathbf{k},\omega)/\phi(\mathbf{k},\omega) = \varepsilon(\mathbf{k},\omega)$, the dielectric function $\varepsilon(\mathbf{k},\omega)$ of the metal consists of the electronic part ε^{el} as well as the ionic part ε^{ion}.

Suppose we regard the bare ions in the metal as well as $\phi^{(0)}$ as external, and the conduction electrons to be the "system" whose response we are studying. Clearly,

$$\phi = \frac{1}{\varepsilon^{\text{el}}}(\phi^{(0)} + \phi^{\text{ion}}_{\text{bare}}). \tag{6.66}$$

Now suppose we regard the electron density in the metal as well as $\phi^{(0)}$ as external, and the bare ions the "system" whose response we are studying. Then we can say

$$\phi = \frac{1}{\varepsilon^{\text{ion}}_{\text{bare}}}(\phi^{(0)} + \phi^{\text{el}}). \tag{6.67}$$

In addition, the superposition principle tells us that

$$\phi = \phi^{(0)} + \phi^{\text{el}} + \phi^{\text{ion}}_{\text{bare}}. \tag{6.68}$$

Using Eq.(6.66), Eq.(6.67) and Eq.(6.68) along with the definition $\phi = (1/\varepsilon)\phi^{(0)}$, we obtain a relation between the above dielectric functions:

$$\varepsilon = \varepsilon^{\text{el}} + \varepsilon^{\text{ion}}_{\text{bare}} - 1. \tag{6.69}$$

There is another way of viewing the dielectric response of a metal. The conduction electrons, in addition to screening the external potential, also screen (dress) the Coulomb potential of the ions. For these dressed ions, the external potential is $\phi_{\text{dressed}} = \phi^{(0)}/\varepsilon^{\text{el}}$. Denoting the dielectric function of these dressed ions as $\varepsilon^{\text{ion}}_{\text{dressed}}$, we may write

$$\phi(\mathbf{k},\omega) = \frac{\phi_{\text{dressed}}(\mathbf{k},\omega)}{\varepsilon^{\text{ion}}_{\text{dressed}}} = \frac{\phi^{(0)}(\mathbf{k},\omega)}{\varepsilon^{\text{ion}}_{\text{dressed}}\varepsilon^{\text{el}}} \equiv \frac{\phi^{(0)}(\mathbf{k},\omega)}{\varepsilon(\mathbf{k},\omega)}. \tag{6.70}$$

[12] Prior to the lattice distortion, the Brillouin zone boundaries were at $k = \pm\pi/a$.

It is not difficult to obtain the dielectric response of the bare ions. Since the ions are heavy and move slowly, we can describe their response classically. We introduce a driving electric field $\mathbf{E}_{\rm loc}e^{-i\omega t}$ to the equation of motion of the ions given by Eq.(6.60):

$$M\ddot{\mathbf{x}} + \frac{n_i(Ze)^2}{\epsilon_0}\mathbf{x} = (Ze)\mathbf{E}_{\rm loc}e^{-i\omega t}.$$

The steady state solution is

$$\mathbf{x} = \frac{\mathbf{E}_{\rm loc}(Ze)/M}{\Omega_p^2 - \omega^2}e^{-i\omega t},$$

and the polarization response is

$$\mathbf{P}e^{-i\omega t} = n_i(Ze)\mathbf{x} = \frac{\epsilon_0 \mathbf{E}_{\rm loc}\Omega_p^2}{\Omega_p^2 - \omega^2}e^{-i\omega t}. \tag{6.71}$$

Now $\mathbf{E}_{\rm loc}$ is the magnitude of the driving field acting on the ions but it is *not* the effective field $\mathbf{E}_0$ prevailing in the system of ions. Consider a slab which has a polarization $\mathbf{P}$ in a direction normal to the slab. The local electric field $\mathbf{E}_{\rm loc}$ acting on a sheet of charge in the slab is found by removing the sheet while keeping the other charges fixed, and calculating the field of these remaining charges. It is easily seen that this local field is given by

$$\mathbf{E}_{\rm loc} = \mathbf{E}_0 + \frac{\mathbf{P}}{\epsilon_0}.$$

Using $\mathbf{P} = \epsilon_0(\varepsilon^{\rm ion}_{\rm bare} - 1)\mathbf{E}_0$, we have

$$\mathbf{P} = \epsilon_0(1 - 1/\varepsilon^{\rm ion}_{\rm bare})\mathbf{E}_{\rm loc}. \tag{6.72}$$

Combining Eq.(6.71) and Eq.(6.72) we get the dielectric response for the bare ions:

$$\varepsilon^{\rm ion}_{\rm bare} = 1 - \frac{\Omega_p^2}{\omega^2}. \tag{6.73}$$

We are now in a position to obtain the dielectric function ε of the metal. For small wavevectors, we can use the Thomas-Fermi value for the dielectric function of the electrons, $\varepsilon^{\rm el}$. Then using Eq.(6.73) in Eq.(6.70) we get

$$\varepsilon(\mathbf{k},\omega) = 1 + \frac{1}{(\lambda_{TF}\mathbf{k})^2} - \frac{\Omega_p^2}{\omega^2} \equiv \left(1 + \frac{1}{(\lambda_{TF}\mathbf{k})^2}\right)\left(1 - \frac{\omega(\mathbf{k})^2}{\omega^2}\right), \tag{6.74}$$

where $\omega(\mathbf{k})$ is the Bohm-Staver phonon frequency obtained in Eq.(6.63). The screened Coulomb interaction energy, $U(\mathbf{k},\omega) = (e^2/\epsilon_0)/k^2\varepsilon(\mathbf{k},\omega)$, of two point charges in a metal is therefore

$$U(\mathbf{k},\omega) = \frac{(e^2/\epsilon_0)}{\mathbf{k}^2 + (1/\lambda_{TF})^2}\left(1 + \frac{\omega(\mathbf{k})^2}{\omega^2 - \omega(\mathbf{k})^2}\right). \tag{6.75}$$

The frequency dependence shows that the effective Coulomb interaction in a metal is not instantaneous. This is a reflection of the fact that unlike the fast electrons, the comparatively heavy ions are much slower in responding to an external potential. Note that for $|\omega| < |\omega(\mathbf{k})|$, we have $U(\mathbf{k},\omega) < 0$ so that the effective Coulomb interaction is attractive. For $|\omega| > |\omega(\mathbf{k})|$, the effective Coulomb interaction is repulsive. If two electrons with energies differing by ω and momenta differing by $\mathbf{k}$ interact in a metal, their effective interaction is attractive as long as $|\omega| < |\omega(\mathbf{k})|$. For example, this condition can be easily satisfied if we consider two electrons at diametrically opposite ends of the Fermi surface and with energies very close to each other. Such pairs of effectively attracting electrons eventually give rise to the phenomenon of superconductivity. As the maximum phonon frequency is the Debye frequency ω_D, electrons with energy differing by more than $\hbar\omega_D$ will not be able to have an effectively attractive Coulomb interaction.

6.4.2.3 Phonon correction to electron energies

The difference between the dielectric function of a free electron gas and an electron gas in a metal is that in a metal, the motion of the positively charged ions in the lattice also makes a physically significant contribution, especially at frequencies smaller than the Debye frequency. The effect of these phonons on the dielectric function of the metal should eventually show up in the potential energy of the electron.

Let us estimate the Hartree-Fock exchange energy $E_{\mathbf{k}}^{\text{exch}}$ using the screened Coulomb potential $U(\mathbf{k}-\mathbf{k}', E_{\mathbf{k}}-E_{\mathbf{k}'})$ of Eq.(6.75) instead of the bare Coulomb potential we used earlier in Eq.(6.43). We have

$$E_{\mathbf{k}}^{\text{exch}} = -\frac{e^2}{\epsilon_0}\int_{k'<k_F}(d\mathbf{k}')\frac{1}{(\mathbf{k}-\mathbf{k}')^2+(1/\lambda_{TF})^2} \times\left(1+\frac{\omega(\mathbf{k}-\mathbf{k}')^2}{[(E_{\mathbf{k}}-E_{\mathbf{k}'})/\hbar]^2-\omega(\mathbf{k}-\mathbf{k}')^2}\right). \quad (6.76)$$

Since the $E_{\mathbf{k}} = \epsilon_{\mathbf{k}}^{(0)} + E_{\mathbf{k}}^{\text{exch}}$ depend on $E_{\mathbf{k}}^{\text{exch}}$, it is quite difficult to solve the above integral equation for the exchange energy. Nevertheless, it is possible to extract useful information, and also to make some simplifications.

We note in Eq.(6.76) that when the phonon contribution is negligible, we recover the exchange energy corresponding to the Thomas-Fermi potential (see Eq.(6.45)):

$$E_{\mathbf{k}}^{TF} = \epsilon_{\mathbf{k}}^{(0)} - \frac{e^2}{\epsilon_0}\int_{k'<k_F}(d\mathbf{k}')\frac{1}{(\mathbf{k}-\mathbf{k}')^2+(1/\lambda_{TF})^2}.$$

The phonon contribution is important when the difference in energies of the two electrons, $(E_{\mathbf{k}}-E_{\mathbf{k}'})$, is close to the phonon energy, $\hbar\omega(\mathbf{k}-\mathbf{k}')$. If the electron energies differ by more than the Debye frequency, ω_D, then the phonon contribution is clearly small. Since ω_D is typically much smaller than the characteristic scale $\epsilon_F/\hbar$ over which the electronic energies vary, a relatively thin shell of electron energies $E_{\mathbf{k}'}$ within ω_D from $E_{\mathbf{k}}$ affect the phonon contribution.

The integral over $\mathbf{k}'$ can be transformed into an integral over the surface $S: E = E_{\mathbf{k}'}$ and the "radial" direction:

$$d\mathbf{k}' = \frac{dS\,dE}{|dE/d\mathbf{k}'|}.$$

Completing the integration over the "radial" variable E between the limits 0 and the Fermi energy ϵ_F leads us to

$$E_{\mathbf{k}} = E_{\mathbf{k}}^{TF} - \frac{e^2}{16\pi^3\epsilon_0}\int_{E_{\mathbf{k}'}=E_{\mathbf{k}}} \frac{dS}{|dE/d\mathbf{k}'|}\frac{1}{(\mathbf{k}-\mathbf{k}')^2+(1/\lambda_{TF})^2}\hbar\omega(\mathbf{k}-\mathbf{k}') \times \ln\left|\frac{\epsilon_F - E_{\mathbf{k}} - \hbar\omega(\mathbf{k}-\mathbf{k}')}{\epsilon_F - E_{\mathbf{k}} + \hbar\omega(\mathbf{k}-\mathbf{k}')}\right|. \tag{6.77}$$

The first term in Eq.(6.77) is the electron energy in the Thomas-Fermi approximation, and the second part is due to phonons. Many conclusions follow from Eq.(6.77).

First, if we substitute $E_{\mathbf{k}} = \epsilon_F$, we find that the phonon part vanishes and we get $\epsilon_F = E_{\mathbf{k}}^{TF}$. Thus the Fermi energy is affected by electron-electron interaction but *unaffected* by coupling of the electrons with phonons. Second, if $|\epsilon_F - E_{\mathbf{k}}| \gg \omega_D$, the logarithm in the phonon part is again small, and $E_{\mathbf{k}} \approx E_{\mathbf{k}}^{TF}$. Thus phonons do not affect electron energies that are far from the Fermi surface. Finally, consider $|E_{\mathbf{k}} - \epsilon_F| \ll \omega_D$. The logarithm is approximately equal to $2(E_{\mathbf{k}} - \epsilon_F)/\hbar\omega(\mathbf{k}-\mathbf{k}')$, so Eq.(6.77) becomes

$$\begin{aligned} E_{\mathbf{k}} &\approx E_{\mathbf{k}}^{TF} - \frac{e^2(E_{\mathbf{k}}-\epsilon_F)}{8\pi^3\epsilon_0}\int_{E_{\mathbf{k}'}=\epsilon_F} \frac{dS}{|dE/d\mathbf{k}'|}\frac{1}{(\mathbf{k}-\mathbf{k}')^2+(1/\lambda_{TF})^2} \\ &\equiv E_{\mathbf{k}}^{TF} + (E_{\mathbf{k}} - \epsilon_F)\lambda, \end{aligned} \tag{6.78}$$

where

$$\lambda = \frac{e^2}{8\pi^3\epsilon_0}\int_{E_{\mathbf{k}'}=\epsilon_F} \frac{dS}{|dE/d\mathbf{k}'|}\frac{1}{(\mathbf{k}-\mathbf{k}')^2+(1/\lambda_{TF})^2}. \tag{6.79}$$

We can obtain a bound on λ :

$$\lambda \le \frac{e^2\lambda_{TF}^2}{8\pi^3\epsilon_0}\int_{E_{\mathbf{k}'}=\epsilon_F} \frac{dS}{|dE/d\mathbf{k}'|}. \tag{6.80}$$

Since $\lambda_{TF}^2 = e^2\nu(0)/\epsilon_0$, and the density of states $\nu(0)$ can also be expressed as (see Eq.(6.57))

$$\nu(0) = \frac{1}{4\pi^3}\int_{E_{\mathbf{k}'}=\epsilon_F} \frac{dS}{|dE/d\mathbf{k}'|},$$

we have

$$\lambda \lesssim 1. \tag{6.81}$$

Rearranging Eq.(6.78), we get the phonon-modified electron energies:

$$E_{\mathbf{k}} - \epsilon_F = \frac{E_{\mathbf{k}}^{TF} - \epsilon_F}{1+\lambda}, \qquad |E_{\mathbf{k}} - \epsilon_F| \ll \omega_D. \tag{6.82}$$

As λ is of the order of unity, this renormalization of electron energies near the Fermi surface by electron-phonon interactions is a significant effect. We have already mentioned at the end of our discussion on Hartree-Fock that the electron energies do not differ much from the free electron value when one uses the Thomas-Fermi (screened) Coulomb interaction to estimate the exchange energy. Thus the deviation of the dispersion relation from the free electron value (in the vicinity of the Fermi surface) is largely due to electron-phonon coupling and not due to electron-electron interaction. This is the case in a large number of metals.

6.5 Anderson orthogonality catastrophe

In the phenomenological development of the Landau Fermi liquid theory, one begins with a gas of free fermions and the interaction is turned on very slowly (quasi-statically) so that the non-interacting energy eigenvalues $\epsilon_{\mathbf{k}}^{(0)}$ evolve to the interacting eigenvalues $\epsilon_{\mathbf{k}}$. The momentum $\mathbf{k}$ is a good quantum number even for the interacting system. If we begin with the ground state Ψ_0 of the free fermion gas, then turning on the interaction quasi-statically takes us to the ground state Ψ of the interacting system. If however a perturbation is turned on suddenly, $V(\mathbf{r},t) = V_0(\mathbf{r})\theta(t)$, then in general the final state will not be the ground state. Such a situation can arise, for example, when an electron is ejected from a core shell of an atom in a metal by X-rays.

The probability P that a system with N fermions still remains in the ground state after the perturbation is turned on is given by the absolute square of their overlap,

$$P = \left| \int d\mathbf{r}_1 \cdots d\mathbf{r}_N \Psi_0^* \Psi \right|^2.$$

For example, if after the introduction of the perturbation, there is a bound state formed, then the overlap of the bound state with the original corresponding free state would be of the order of $N^{-1/2}$, where N is the number of fermions. Anderson showed [13] that, in general, the overlap of the ground

[13] P. W. Anderson, *Phys. Rev. Lett.* **18**, 1049 (1967).

state (many-body) wavefunctions of a gas of N fermions before and after the perturbation will behave as $N^{-\alpha}$ where $\alpha > 0$. Thus for large N, the overlap approaches zero. It is in this sense that the ground states are "orthogonal", and hence, we have an *orthogonality catastrophe.* We shall now estimate this overlap.

Consider a large spherical cavity of radius R and with an infinite wall. For $t < 0$ the perturbing potential is absent and the $l = 0$ wavefunctions of the Schrödinger equation are given by

$$\varphi_n = C_n \frac{\sin(k_n r)}{k_n r}, \qquad k_n R = n\pi,\ \epsilon_n = \frac{\hbar^2 k_n^2}{2m} < \epsilon_F. \tag{6.83}$$

At $t > 0$, a short-range[14] perturbation $V_0(r)$ acts on the system, and we assume $V_0(r)$ is centered about the origin. The new $l = 0$ wavefunctions are

$$\psi_n = C'_n \frac{\sin(k'_n r + \delta_0(\epsilon'_n))}{k'_n r}, \qquad k'_n R + \delta_0 = n\pi = k_n R. \tag{6.84}$$

The overlap of the wavefunctions belonging to the two sets is (for $\delta \neq 0$)

$$\begin{aligned} A_{n_1 n_2} &= 4\pi C_{n_1} C'_{n_2} \int_0^R dr\, r^2 \frac{\sin(k_{n_1} r)\sin(k'_{n_2} r + \delta(\epsilon'_{n_2}))}{(k_{n_1} r)(k'_{n_2} r)} \\ &= \frac{2\pi C_{n_1} C'_{n_2} \sin(\delta(\epsilon'_{n_2}))}{k_{n_1}(k_{n_2} - \delta(\epsilon'_{n_2})/R)} \left[\frac{1}{k_{n_1} - k_{n_2} + \delta(\epsilon'_{n_2})/R} \right. \\ &\qquad \left. + \frac{1}{k_{n_1} + k_{n_2} - \delta(\epsilon'_{n_2})/R} \right]. \end{aligned} \tag{6.85}$$

Using the normalizations

$$C_n = \frac{k_n}{\sqrt{2\pi R}}, \text{ and } C'_n = \frac{k'_n}{\sqrt{2\pi R}},$$

the coefficients $A_{n_1 n_2}$ become

$$A_{n_1 n_2} = \sin(\delta(\epsilon'_{n_2})) \left[\frac{1}{\pi(n_1 - n_2) + \delta(\epsilon'_{n_2})} + \frac{1}{\pi(n_1 + n_2) - \delta(\epsilon'_{n_2})} \right]. \tag{6.86}$$

For $n_1 = n_2 = n$, the overlap is approximately unity for small phase shifts:

$$A_{nn} \approx \frac{\sin(\delta(\epsilon'_n))}{\delta(\epsilon'_n)} \approx 1,\ \delta(\epsilon'_n) \ll 1.$$

Further, if $n_1 \neq n_2$, for small phase shifts, the overlap approaches zero. Thus for a weak perturbation, the final states have a very good overlap with the corresponding unperturbed states.

[14]See Chapter 4 on Scattering Theory for a definition of short range.

The ground state wavefunctions in the Hartree-Fock approximation in the two cases are, respectively,

$$\Psi_0(\mathbf{r}_1,\cdots,\mathbf{r}_n) = P_A \prod_{n,\epsilon_n\le\epsilon_F} \varphi_n(\mathbf{r}_n), \qquad \Psi(\mathbf{r}_1,\cdots,\mathbf{r}_n) = P_A \prod_{n,\epsilon'_n\le\epsilon_F} \psi_n(\mathbf{r}_n), \tag{6.87}$$

where P_A is the antisymmetric sum ensuring that the wavefunctions satisfy the Pauli principle. The overlap $\int\prod_i d\mathbf{r}_i\Psi_0^*\Psi$ is then the determinant of the matrix A whose elements $A_{n_1n_2}$are given by Eq.(6.86) subject to the requirement $\epsilon_{n_1},\epsilon'_{n_2}\le\epsilon_F$ because the ground state wavefunction has occupied states with energy up to a maximum value of ϵ_F. So we have

$$\begin{aligned}P &= (\det[A_{n_1n_2}])(\det[A_{n_2n_1}]),\ \epsilon_{n_1},\epsilon'_{n_2}\le\epsilon_F\\ &= e^{\mathrm{tr}\ln AA^T}.\end{aligned} \tag{6.88}$$

Here we used $\det[A] = e^{\ln\det[A]} = e^{\mathrm{tr}\ln A}$. We remark here that $A^T\neq A$. This is clear from Eq.(6.86). We need some more simplifications before we can estimate P. Consider the expansion of the nth perturbed wavefunction in terms of the unperturbed wavefunctions:

$$\psi_{n_1} = \sum_{n_2,\epsilon_{n_2}\le\epsilon_F} A_{n_2n_1}\varphi_{n_2} + \sum_{n_2,\epsilon_{n_2}>\epsilon_F} A_{n_2n_1}\varphi_{n_2}. \tag{6.89}$$

Since the ψ_n are normalized, the coefficients $A_{n_2n_1}$ will satisfy

$$\sum_{n_2} A_{n_2n_1}A_{n_2n_3} = \sum_{n_2} A^T_{n_1n_2}A_{n_2n_3} = \delta_{n_1n_3}. \tag{6.90}$$

Eq.(6.90) implies

$$\sum_{n_2,\epsilon_{n_2}\le\epsilon_F} A^T_{n_1n_2}A_{n_2n_3} = \delta_{n_1n_3} - \sum_{n_2,\epsilon_{n_2}>\epsilon_F} A^T_{n_1n_2}A_{n_2n_3}. \tag{6.91}$$

By substituting Eq.(6.91) in Eq.(6.88), the probability P for remaining in the ground state can be shown to be

$$\begin{aligned}P &= \exp\left[\mathrm{tr}\left(\ln\left[\delta_{n_1n_3} - \sum_{n_2,\epsilon_{n_2}>\epsilon_F} A^T_{n_1n_2}A_{n_2n_3}\right]\right)\right]\\ &\lesssim \exp\left[-\sum_{n_1,\epsilon'_{n_1}\le\epsilon_F}\ \sum_{n_2,\epsilon_{n_2}>\epsilon_F} A^T_{n_1n_2}A_{n_2n_1}\right],\end{aligned} \tag{6.92}$$

which leads to

$$P \lesssim \exp\left[-\sum_{n_1,\epsilon'_{n_1}\le\epsilon_F}\ \sum_{n_2,\epsilon_{n_2}>\epsilon_F} \frac{\sin^2(\delta(\epsilon'_{n_1}))}{[(n_2-n_1)\pi+\delta(\epsilon'_{n_1})]^2}\right]. \tag{6.93}$$

We have dropped the second contribution, $\sin(\delta)/[(n_2+n_1)\pi-\delta]$, in the overlap $A_{n_2 n_1}$ because the large denominator ($\pi n_2 > k_F R \gg 1$) makes this term much less significant than the first. The sum in Eq.(6.93) is dominated by values of n_1 and n_2 close to $k_F/(\pi R)$. The sum over n_1 runs from 1 to $n_F = k_F R/\pi$. The sum over n_2 runs from n_F+1 to a large value $D \gg n_F$. We approximate the sums by integrals and perform the n_2 integral first:

$$P \lesssim \exp\left[\frac{1}{\pi^2}\int_1^{n_F=k_F R/\pi} dn_1 \frac{\sin^2(\delta(\epsilon'_{n_1}))}{n_F+1-n_1+\delta(\epsilon'_{n_1})/\pi}\right]$$
$$\approx \exp\left[-\frac{\sin^2(\delta(\epsilon_F))}{\pi^2}\ln\left(\frac{k_F R}{\pi}\right)\right]. \tag{6.94}$$

Making use of $k_F = (3\pi^2 N/V)^{1/3}$ in Eq.(6.94), the probability to remain in the ground state can be written as

$$P \lesssim N^{-\sin^2(\delta(\epsilon_F))/3\pi^2}. \tag{6.95}$$

Thus in the limit of a large number of fermions, $P \to 0$ even if $\delta(\epsilon_F)$ happens to be small. A more accurate estimate of P was made in a subsequent paper by Anderson. The result is

$$P \sim N^{-\delta(\epsilon_F)^2/\pi^2}. \tag{6.96}$$

When the potential forms one bound state, $\delta(\epsilon_F)$ for spinless fermions is π, and Eq.(6.96) yields $P \sim 1/N$ as expected. Had we taken into account the spin of the fermions, then, say, for $S = 1/2$, the exponents in Eq.(6.95) and Eq.(6.96) would have been enhanced by a factor of two:

$$P \simeq N^{-2\delta(\epsilon_F)^2/\pi^2} = N^{-(\Delta N_\uparrow)^2-(\Delta N_\downarrow)^2}. \qquad \text{(including spin)} \tag{6.97}$$

In the last step, we identified δ/π as the change in the number of bound states for each spin species (Friedel sum rule).

If one begins with the ground state of a system of (spinless) electrons in a perturbing potential, and the perturbation undergoes a change, then the probability to remain in the ground state is

$$P \sim N^{-(\delta_f(\epsilon_F)-\delta_i(\epsilon_F))^2/\pi^2}, \tag{6.98}$$

where δ_i and δ_f are the initial and final phase shifts, respectively.

6.6 Diffusion of muons in metal crystals

The diffusion of muons or protons in metals at low temperature provides an interesting example of the Anderson orthogonality catastrophe. The muons

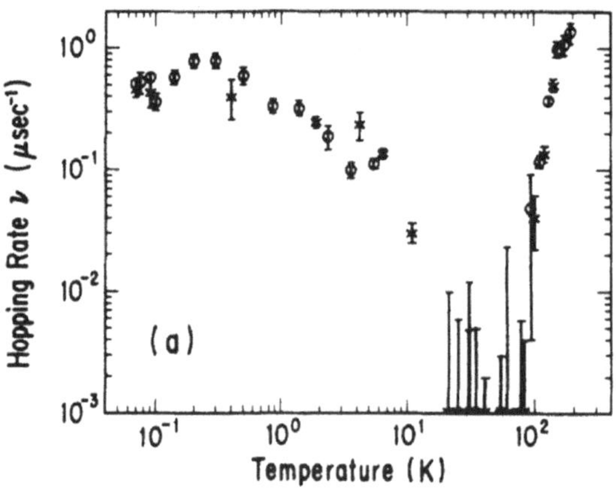

FIGURE 6.9: Data showing the muon hopping rate ν in copper as a function of temperature. The data between about 15K and 0.1K is well described by the law $\nu \propto (k_B T/\epsilon_F)^{2K-1}$, $K \approx 0.2$. At the lowest temperatures, the rate tends to saturate and deviates from the predicted power law. Source: R. Kadono *et al.*, *Phys. Rev.* B **39**, 23 (1989).

(and protons) are positively charged, so the natural position for them to occupy is the interstitial sites in the crystal. The interstitial sites may themselves form a lattice. The motion of the muons from one site to the other is impeded due to many causes. First, there is the potential barrier due to the lattice ions. Second, a muon sitting at an interstitial site causes a local distortion of the lattice by repelling the positively charged lattice ions around it. This distortion needs to be carried along when the muon moves. Finally, the hopping muon creates an abrupt change of potential for the conduction electrons, so coherent hopping will be additionally suppressed because of the non-perfect overlap of the electron state before and after the hopping. We are interested in this last cause.

The physical effect was predicted by Kondo[15] as well as Yamada[16] and coworkers. The temperature dependence of the muon diffusion rate is predicted to be $\nu \sim (D_0/a^2)(k_B T/\epsilon_F)^{2K-1}$, where D_0 is a diffusion constant, a is the muon hopping distance, and $K \leq 1/2$. This is in good agreement with data for the muon hopping rate in copper[17] as can be seen in Fig. 6.9.

Experimentally, it is better to study muon diffusion instead of proton diffusion because the muons are much lighter, and therefore, have a much higher tunneling probability between the interstitial sites.

---

[15] J. Kondo, *Physica* 84**B**, 40 (1976)

[16] K. Yamada and K. Yosida, *Prog. Theor. Phys.* **68**, 1504 (1982); **59**, 1061 (1978); **60**, 353 (1978); **62**, 363 (1979).

[17] See R. Kadono *et al. Phys. Rev.* B **39**, 23 (1989).

When the muon hops from an interstitial site at a position $\mathbf{r}_1$ to another at $\mathbf{r}_2$, its electrostatic potential $V(\mathbf{r}-\mathbf{r}_1)$ centered at $\mathbf{r}_1$ disappears and the same potential then reappears centered at $\mathbf{r}_2$ as $V(\mathbf{r}-\mathbf{r}_2)$. Let δ_i and δ_f be the phase shifts, respectively, before and after the hopping. Unlike the previous discussion on Anderson orthogonality, these phase shifts correspond to the annihilation and creation of the perturbation at *different* sites. It turns out that if the two interstitial site positions are very far apart compared to the Fermi wavelength ($a = |\mathbf{r}_2 - \mathbf{r}_1| \gg k_F^{-1}$), then the overlap of the electronic ground state before and after the hopping may be regarded as the product of two independent overlaps at the two sites:

$$\left|\int \prod_n d\mathbf{r}_n \Psi_i^* \Psi_f\right| \sim N^{-(\delta_i^2+\delta_f^2)/2\pi^2} = N^{-\delta^2/\pi^2}. \qquad \text{(no spin, } l=0\text{)} \quad (6.99)$$

Here we assumed for simplicity that the two interstitial sites have the same environment, so $\delta_i = \delta_f = \delta$. More generally, Yamada has shown that for $l = 0$ spinless fermions,

$$\left|\int \prod_n d\mathbf{r}_n \Psi_0^* \Psi\right| \sim N^{-K}, \qquad (6.100)$$

where K is related to the phase shift δ and the spherical Bessel function $x = j_0^2(k_F a) = \sin^2(k_F a)/(k_F a)^2$:

$$\begin{aligned} K &= K_0(x,\delta) \\ &= \begin{cases} \left[\frac{1}{\pi}\tan^{-1}\left(\frac{\sqrt{1-x}}{\sqrt{1+x\tan^2\delta}}\tan\delta\right)\right]^2, & 0 \le \delta < \pi/2 \\ \frac{2}{\pi}\tan^{-1}\sqrt{\frac{1-x}{x}} - \left[\frac{1}{\pi}\tan^{-1}\left(\frac{\sqrt{1-x}}{\sqrt{1+x\tan^2\delta}}|\tan\delta|\right)\right]^2, & \pi/2 \le \delta < \pi. \end{cases} \end{aligned} \qquad (6.101)$$

Since $x \to 0$ as $k_F a \to \infty$, we get the expected result for large hopping distances in agreement with Eq.(6.99):

$$K = K(0,\delta) = \frac{\delta^2}{\pi^2}, \quad 0 \le \delta < \pi. \qquad \text{(distant hopping)} \qquad (6.102)$$

Zero hopping distance, or $x = 1$, corresponds to the muon remaining where it was. Not surprisingly, the overlap is complete:

$$K = K(1,\delta) = 0. \qquad \text{(no hopping).} \qquad (6.103)$$

If the electron spin is also considered, and the scattering of electrons from the muon (and the screening of its charge) is dominated by the angular momentum channel $l = l_0$, then the phase shift δ and the coefficient K will be

$$\begin{aligned} \delta &= \frac{\pi}{2(2l_0+1)}, \\ K &= 2(2l_0+1)K_0(x,\delta). \qquad \text{(spinful case)} \end{aligned} \qquad (6.104)$$

The first equation is just a statement of the fact that the screening cloud has a unit total charge (Friedel sum rule). The maximum value of K in the spinful case is easily seen to be

$$K^{\max} = \frac{1}{2(2l_0+1)}. \qquad \text{(with spin)} \tag{6.105}$$

We now discuss the temperature dependence of the muon hopping due to Anderson orthogonality. When the muon hops, the number N of quasiparticles near the Fermi surface whose phases are affected is of the order of $1/(\nu(0)Vk_BT) \approx \epsilon_F/k_BT$. If t is the bare hopping amplitude for the muon between neighboring interstitial sites at a distance a, then the dressed hopping amplitude $\tilde{t}$ taking into account the overlap of the quasiparticle clouds is

$$\tilde{t} = tN^{-K} = \left(\frac{k_BT}{\epsilon_F}\right)^K t. \tag{6.106}$$

Finally we estimate the muon diffusion constant D_μ. The diffusion arises because of the scattering the muon undergoes with the conduction electrons. If $T(\mathbf{k},\mathbf{k}')$ is the $T-$matrix for muon-electron scattering, the scattering rate $\tau_{\mu e}$ in the relaxation time approximation is given by

$$\begin{aligned}\frac{1}{\tau_{\mu e}} &= \frac{2\pi}{\hbar}\sum_{\mathbf{k},\mathbf{k}'}|T(\mathbf{k},\mathbf{k}')|^2\delta(\epsilon_{\mathbf{k}}-\epsilon_{\mathbf{k}'})n(\epsilon_{\mathbf{k}})(1-n(\epsilon_{\mathbf{k}'}))\\ &\approx \frac{2\pi}{\hbar}|T_{av}(\mathbf{k},\mathbf{k}')\nu(0)|^2(k_BT).\end{aligned}$$

The muon diffusion constant is

$$D_\mu \approx v_\mu^2\tau_{\mu e} \approx \frac{\tilde{t}^2a^2}{\hbar^2}\tau_{\mu e} = \left(\frac{k_BT}{\epsilon_F}\right)^{2K-1} D_0. \tag{6.107}$$

Here $v_\mu \approx \tilde{t}a/\hbar$ is the velocity of the muon. The muon hopping rate is $\nu = D_\mu/a^2$. Since the maximum value of K is $1/2$, the muon diffusion constant and hopping rate *increase* as the temperature is reduced. This increase does not continue indefinitely because the muon has a finite lifetime. The muon decay rate is about $\gamma_\mu = 5\times 10^5\mathrm{s}^{-1}$. When the scattering time $\tau_{\mu e}$ exceeds the muon lifetime $1/\gamma_\mu$, which is going to be the case at sufficiently low temperatures, then the muon diffusion stops increasing with further decrease of temperature. This is believed to be the case in copper where the muon diffusion initially increases according to a power law as the temperature is reduced but eventually saturates [see R. Kadono *et al.*, Phys. Rev. B **39**, 23 (1989).]. In aluminium, the saturation is not observed down to about 0.1K possibly because its scattering time $\tau_{\mu e}$ is much smaller than that of copper due to a larger hopping element $\tilde{t}$.

6.7 Screening of transverse fields in metals

The linear response of an isotropic electron gas to external charges and fields is characterized by the response functions - the polarizability $\Pi(\mathbf{q},\omega)$ and the conductivity $\sigma(\mathbf{q},\omega)$:

$$\begin{aligned}\rho_{\text{ind}}(\mathbf{q},\omega) &= \Pi(\mathbf{q},\omega)\phi(\mathbf{q},\omega),\\ \mathbf{j}_{\text{ind}}(\mathbf{q},\omega) &= \sigma(\mathbf{q},\omega)\mathbf{E}(\mathbf{q},\omega).\end{aligned} \tag{6.108}$$

The first of these has already been encountered in Eq. (6.5) in the time-independent version, while the second is Ohm's law. The two response functions are not strictly independent since they are related through the continuity equation, $\partial\rho_{\text{ind}}/\partial t + \nabla\cdot\mathbf{j}_{\text{ind}} = 0$, from which it follows that

$$\Pi(\mathbf{q},\omega) = -i(q^2/\omega)\sigma(\mathbf{q},\omega), \tag{6.109}$$

whence the dielectric function $\epsilon(\mathbf{q},\omega)$ can also be expressed in terms of $\sigma(\mathbf{q},\omega)$:

$$\epsilon(\mathbf{q},\omega) = 1 - \frac{1}{\epsilon_0 q^2}\Pi(\mathbf{q},\omega) \equiv 1 + \frac{i}{\epsilon_0\omega}\sigma(\mathbf{q},\omega). \tag{6.110}$$

Similar to the dielectric function, one also has the corresponding magnetic permittivity μ such that $\mathbf{B}(\mathbf{q},\omega) = \mu_0\mu(\mathbf{q},\omega)\mathbf{H}(\mathbf{q},\omega)$. The Maxwell equations govern the space and time dependence of the electric and magnetic fields in the metal:

$$i\mathbf{q}\cdot\mathbf{E}(\mathbf{q},\omega) = \frac{1}{\epsilon_0}(\rho_{\text{ext}}(\mathbf{q},\omega) + \Pi(\mathbf{q},\omega)\phi(\mathbf{q},\omega)), \tag{6.111}$$

$$\mathbf{q}\times\mathbf{E}(\mathbf{q},\omega) = \omega\mathbf{B}(\mathbf{q},\omega), \tag{6.112}$$

$$\mathbf{q}\cdot\mathbf{B} = 0, \tag{6.113}$$

$$i\mathbf{q}\times\mathbf{H}(\mathbf{q},\omega) = (\mathbf{j}_{\text{ext}} - i\omega\mathbf{P}) - i\omega\epsilon_0\mathbf{E}, \tag{6.114}$$

where we used Eqs. (6.108) for the linear response relations for the induced charge density and the current density. In static (time-independent) problems, as was the case we studied in Sec. 6.1, we found that the electric field resulting from a spatially slowly varying external charge density is screened by the conduction electrons beyond the Thomas-Fermi screening length λ_{TF}.

Now any vector can be resolved into a sum of longitudinal (L) and transverse (T) components, where the longitudinal component is curl-free and the transverse component is divergence-free. In the static case, the electric field is then purely longitudinal. Coming to the time-dependent problem, the electric field has both longitudinal and transverse components, with the latter coming from the second of the above Maxwell equations (Faraday's law). The magnetic field $\mathbf{H}$, being divergence-free, is purely transverse.

The screening of the longitudinal component of the electric field, $\mathbf{E}^L = -\nabla\phi$ has already been discussed earlier in this section. We will now consider the screening of the transverse (magnetic) field. Using the above Maxwell equations, we easily obtain the relation between $\mathbf{H}$ and the external current:

$$\begin{aligned}\mathbf{H} &= -i\frac{\mathbf{q}\times\mathbf{j}_{\text{ext}}}{\omega^2\epsilon\mu/c^2 - q^2} \\ &= -i\frac{\mathbf{q}\times\mathbf{j}_{\text{ext}}}{(\omega^2/c^2)\mu[1+\frac{i}{\epsilon_0\omega}\sigma] - q^2},\end{aligned} \tag{6.115}$$

where $c^2 = 1/(\epsilon_0\mu_0)$. The magnetic field thus gets screened over a length scale $\lambda_{\text{skin}} = 1/\sqrt{\sigma\mu\mu_0\omega}$ which is commonly referred to as the skin depth. The transverse component of the electric field also gets screened over the same length scale. The divergence of the skin depth at low frequencies should be contrasted with the Thomas-Fermi screening length for the longitudinal component of the electric field.

Chapter 7

Collective Excitations

7.1 Plasma waves

In Sec. 6.4.2 we identified phonons in metals as plasma oscillations of the ionic lattice in a medium whose dielectric function is determined by the conduction electrons. Here we address the problem of collective plasma oscillations of the homogeneous electron gas. We have already seen that in the linear response approximation, the electric potential $\phi(\mathbf{q},\omega)$ in a medium subjected to an external potential $\phi^{(0)}(\mathbf{q},\omega)$ is given by $\phi(\mathbf{q},\omega) = \phi^{(0)}(\mathbf{q},\omega)/\epsilon(\mathbf{q},\omega)$. Since collective excitations sustain even when no external source is present (i.e. $\phi^{(0)}(\mathbf{q},\omega) = 0$), the condition for collective charge (or plasma) oscillations of the electron gas is

$$\epsilon(\mathbf{q},\omega) \equiv 1 - \frac{1}{q^2\epsilon_0}\Pi(\mathbf{q},\omega) = 0. \tag{7.1}$$

These plasma oscillations are longitudinal since for our choice of gauge, $\nabla \cdot \nabla A = 0$, the scalar potential generates the longitudinal component of the electric field.

Let us evaluate the polarizability for small wavevectors, $\mathbf{q} \to 0$, in three dimensions in the Lindhard approximation:

$$\begin{aligned}\Pi(\mathbf{q},\omega) &\approx -2e^2 \int (d\mathbf{k}) \frac{\mathbf{q}\cdot\mathbf{v_k}}{\omega - \mathbf{q}\cdot\mathbf{v_k} + i\delta} \frac{\partial n}{\partial \epsilon_{\mathbf{k}}^{(0)}} \\ &= \frac{2qe^2k_F^2}{\hbar(2\pi)^2} \int_0^\pi d\theta \sin\theta \frac{\cos\theta}{\omega - qv_F\cos\theta + i\delta}.\end{aligned} \tag{7.2}$$

Upon completing the integral, we obtain

$$\Pi(\mathbf{q},\omega) \approx -\nu(0)e^2 \left[1 - \frac{\omega}{2v_F q}\ln\left|\frac{\omega + qv_F}{\omega - qv_F}\right| + i\frac{\pi|\omega|}{2v_F q}\Theta(v_F q - |\omega|)\right]. \tag{7.3}$$

Expanding the logarithm in Eq. 7.3 in powers of $qv_F/\omega \ll 1$ (to the fifth order) we get

$$\Pi(\mathbf{q},\omega) \approx \nu(0)v_F^2\left(\frac{q^2}{3\omega^2}\right)\left[1+\frac{3}{5}\left(\frac{v_F q}{\omega}\right)^2\right], \tag{7.4}$$

and substituting this in Eq. 7.1, we obtain the dispersion relation for collective plasma oscillations of the electron gas in three dimensions:

$$\omega^2 \approx \omega_p^2 + \frac{3}{5}(v_F q)^2, \tag{7.5}$$

where $\omega_p = \sqrt{ne^2/m\epsilon_0}$ is the electron plasma frequency. The plasma oscillations are undamped as long as the imaginary component of $\Pi(\mathbf{q},\omega)$ is zero. Damping occurs for $\hbar|\omega| < v_F q$, or more accurately where

$$\hbar\omega = \epsilon^{(0)}_{\mathbf{k}+\mathbf{q}} - \epsilon^{(0)}_{\mathbf{k}} \tag{7.6}$$

is satisfied.

Let us now consider transverse plasma oscillations. The condition for self-sustaining transverse oscillations is given by the vanishing of the denominator in Eq. 6.115,

$$(\omega^2\mu/c^2 - q^2) + i\sigma(\mathbf{q},\omega)\omega\mu\mu_0 = 0. \tag{7.7}$$

Using the relation between σ and Π, (see Eq. 6.109) and the small wavevector expansion for Π in Eq. 7.3, we obtain the dispersion relation for transverse plasma oscillations,

$$\omega^2 = \omega_p^2 + \frac{1}{\mu}(cq)^2. \tag{7.8}$$

7.2 Spin waves

Spin waves – slow spatial variations of the magnetization – are undamped low-energy excitations of a system in a magnetically ordered ground state.

In our earlier study of the Hartree-Fock approximation for the free electron gas, we found that Coulomb interactions together with the requirement of antisymmetry of the many-particle electron wavefunction give rise to an exchange interaction of the electrons, $f(\mathbf{k}\sigma,\mathbf{k}'\sigma') = -\delta_{\sigma,\sigma'}(e^2/\epsilon_0)\times 1/|\mathbf{k}-\mathbf{k}'|^2$, which evidently favors parallel alignment of the electron spins. Within this approximation, we showed in Sec. 6.3.2.2 that at sufficiently low densities, the homogeneous electron gas interacting through the Coulomb potential is unstable against a coincident ferromagnetic and charge density instability, which is evidenced in the divergence of the uniform magnetic susceptibility and compressibility. More microscopically, one considers the Lindhard polarizability (which determines the linear response of an electron gas to small external perturbations, including magnetic) but instead of the bare energies $\epsilon_{\mathbf{k}}^{(0)}$, one uses the modified energies $E_{\mathbf{k}}$ incorporating the exchange correction as in Eq. 6.43. From the poles of the Lindhard function at finite frequencies and wavevectors, one may, in principle, obtain the spin-wave spectrum. A note of caution here is that since within the aforementioned approximation, the ferromagnetic instability is accompanied by a density instability, it is not clear whether such spin-waves are physically very relevant. On the other hand, we shall see later in Sec. 9.6 that upon approaching the ferromagnetic instability while remaining on the paramagnetic side, weakly damped spin-wave excitations – paramagnons – are possible.

Chapter 8

Effect of Disorder on Electron Interactions

Disorder, commonly encountered in the form of random potential scatterers in the metallic host, introduces an additional time scale – the mean free time, τ, between successive collisions. Elastic scattering of electrons from such defects can change the momentum but not the energy of the scattering particle. This leads to diffusive rather than ballistic propagation of the electron wavepackets, with the diffusion constant $D = \pi v_F^2 \tau / d$, where d is the spatial dimensionality of the system. An inhomogeneity in the electron charge density induces a current $\mathbf{j}_{\text{ind}} = -D\nabla\rho_{\text{ind}}$, which, when combined with the continuity equation, leads us to conclude that

$$\rho_{\text{ind}}(\mathbf{q},\omega) = \rho_{\text{ind}}(\mathbf{q},t=0)\frac{1}{D\mathbf{q}^2 - i\omega}. \tag{8.1}$$

In the presence of diffusion, the long wavelength density fluctuations relax on a much longer time scale, $1/D\mathbf{q}^2$, compared to the characteristic time scale $1/v_F q$ of the relaxation of density modes in the clean case (see Eq. 7.3). We eliminate the dependence on the initial (induced) charge distribution by taking the ratio

$$\frac{\rho_{\text{ind}}(\mathbf{q},\omega)}{\rho_{\text{ind}}(\mathbf{q},0)} = \frac{D\mathbf{q}^2}{D\mathbf{q}^2 - i\omega}. \tag{8.2}$$

Correspondingly, the polarizability in the presence of diffusion acquires a significant frequency dependence even where $\omega \ll v_F q$:

$$\Pi(\mathbf{q},\omega) \approx \Pi(\mathbf{q},0)\frac{D\mathbf{q}^2}{D\mathbf{q}^2 - i\omega}, \qquad \omega \ll v_F q, \tag{8.3}$$

where we approximated the frequency dependence of the polarizability on the right hand side by its static value. The above frequency dependence also appears in the effective Coulomb potential,

$$U(\mathbf{q},\omega) = \frac{U^{(0)}(\mathbf{q})}{1 - \Pi(\mathbf{q},\omega)U^{(0)}(\mathbf{q})} \equiv \frac{U^{(0)}(\mathbf{q})}{\epsilon(\mathbf{q},\omega)}. \tag{8.4}$$

Here $U^{(0)}$ refers to the bare Coulomb potential.

8.1 Altshuler-Aronov corrections to density of states

We consider first corrections to the density of states in a perturbative expansion in the electron interaction. The leading order correction Σ_i to quasiparticle energy ϵ_i (reckoned from the Fermi level) consists of the usual Hartree and exchange (Fock) terms that we have encountered earlier:

$$\Sigma_i = -\sum_{\epsilon_j<0} \int d\mathbf{r}\, d\mathbf{r}' e^2 U(\mathbf{r}-\mathbf{r}') \left[\psi_i^*(\mathbf{r})\psi_j^*(\mathbf{r}')\psi_i(\mathbf{r}')\psi_j(\mathbf{r}) \right.$$
$$\left. -\psi_i^*(\mathbf{r})\psi_j^*(\mathbf{r}')\psi_i(\mathbf{r})\psi_j(\mathbf{r}')\right] \equiv \Sigma_i^{\text{exch}} + \Sigma_i^H. \tag{8.5}$$

Here $\psi_i(\mathbf{r})$ refers to the wavefunction of the ith stationary state with energy ϵ_i, and e^2U is the Coulomb interaction energy. The disorder-averaged exchange correction, $\Sigma_\epsilon^{\text{exch}}$, to the quasiparticle energy is what we are interested in (the Hartree term merely causes a constant shift to the quasiparticle energies):

$$\Sigma_\epsilon^{\text{exch}} = \frac{1}{\nu(0)V}\sum_i \left\langle \delta(\epsilon-\epsilon_i)\Sigma_i^{\text{exch}}\right\rangle, \tag{8.6}$$

with V being the system volume and the angular brackets denoting disorder averaging. Combining Eqs. (8.5) and (8.6), the exchange part of the disorder averaged self energy can be presented as

$$\Sigma_\epsilon^{\text{exch}} = -\frac{e^2}{\nu(0)V}\int d\mathbf{r}\, d\mathbf{r}' U(\mathbf{r}-\mathbf{r}')$$
$$\times \sum_{ij}\int_\epsilon^\infty d\omega \left\langle \delta(\epsilon-\epsilon_i)\delta(\epsilon-\omega-\epsilon_j)\psi_i^*(\mathbf{r})\psi_j^*(\mathbf{r}')\psi_i(\mathbf{r}')\psi_j(\mathbf{r})\right\rangle. \tag{8.7}$$

The second delta-function factor, $\delta(\epsilon-\omega-\epsilon_j)$, with $\omega>\epsilon$, imposes the constraint that the energies ϵ_j are negative, as in Eq. (8.5). We now introduce the quantity

$$F(\epsilon,\omega;\mathbf{r}-\mathbf{r}') = \sum_{ij}\left\langle \delta(\epsilon-\epsilon_i)\delta(\epsilon-\omega-\epsilon_j)\psi_i^*(\mathbf{r})\psi_j^*(\mathbf{r}')\psi_i(\mathbf{r}')\psi_j(\mathbf{r})\right\rangle, \tag{8.8}$$

where the $\mathbf{r}-\mathbf{r}'$ dependence in the LHS reflects the fact that under disorder averaging, translational invariance in space gets restored. Thus,

$$\Sigma_\epsilon^{\text{exch}} = -\frac{e^2}{\nu(0)V}\int_\epsilon^\infty d\omega \int d\mathbf{r}\, d\mathbf{r}' U(\mathbf{r}-\mathbf{r}')F(\epsilon,\omega;\mathbf{r}-\mathbf{r}'). \tag{8.9}$$

It turns out that F is related to the imaginary part of the polarizability function Π. Consider the imaginary part of the polarizability function at zero temperature and energy $\omega > 0$:

$$\mathrm{Im}\Pi(\mathbf{q},\omega) = 2e^2 \int d\mathbf{r}\, d\mathbf{r}' e^{i\mathbf{q}\cdot(\mathbf{r}-\mathbf{r}')} \sum_{ij} \langle \delta(\epsilon_i - \epsilon_j - \omega) \psi_i^*(\mathbf{r}) \psi_j^*(\mathbf{r}') \psi_i(\mathbf{r}') \psi_j(\mathbf{r}) \rangle \tag{8.10}$$

with i referring to unoccupied states ($\epsilon_i > 0$) and j referring to occupied states ($\epsilon_j < 0$). In the absence of disorder, Eq.(8.10) agrees with the frequency dependent polarizability function obtained in Sec. 6.1.3.1. The sum over energies in Eq.(8.10) is easily presented as an integral over energy by introducing our quantity F,

$$\begin{aligned} \mathrm{Im}\Pi(\mathbf{q},\omega) &= 2e^2 \int_0^{\omega} d\epsilon\, d\mathbf{r} F(\epsilon,\omega;\mathbf{r}) e^{i\mathbf{q}\cdot\mathbf{r}} \\ &= 2e^2\omega \int d\mathbf{r} F(\omega;\mathbf{r}) e^{i\mathbf{q}\cdot\mathbf{r}}, \end{aligned} \tag{8.11}$$

where we made use of the fact that upon averaging, F is dependent only on ω but not on ϵ. Combining Eqs. (8.9) and (8.11) together with Eq.(8.3) we finally arrive at the following expression for the exchange correction to the quasiparticle energy:

$$\Sigma_\epsilon^{\mathrm{exch}} = -\frac{e^2}{2\pi} \int_\epsilon^\infty d\omega \int (d\mathbf{q}) U(\mathbf{q}) \frac{D\mathbf{q}^2}{\omega^2 + (D\mathbf{q}^2)^2}. \tag{8.12}$$

To estimate the correction to the density of quasiparticle states near the Fermi level, $\nu(\epsilon)$, we begin with the definition and perform an expansion to leading order in the correction (Σ):

$$\begin{aligned} \nu(\epsilon) &= \frac{1}{V} \sum_i \langle \delta(\epsilon - \epsilon_i - \Sigma_i) \rangle \approx \frac{1}{V} \sum_i \left[\langle \delta(\epsilon - \epsilon_i) \rangle - \langle \Sigma_i \delta'(\epsilon - \epsilon_i) \rangle \right] \\ &\approx \nu(0) \left[1 - \frac{\partial \Sigma_\epsilon^{\mathrm{exch}}}{\partial \epsilon} \right]. \end{aligned} \tag{8.13}$$

We made use of the fact that in the leading order in Coulomb interaction, only the exchange part of the interaction energy depends on quasiparticle energy. Using Eq.(8.12) and Eq.(8.13) we obtain the correction $\delta\nu(\epsilon) = \nu(\epsilon) - \nu(0)$ to the density of states near the Fermi level,

$$\frac{\delta\nu(\epsilon)}{\nu(0)} = -\frac{\partial \Sigma_\epsilon^{\mathrm{exch}}}{\partial \epsilon} = -\frac{e^2}{2\pi} \int (d\mathbf{q}) U(\mathbf{q}) \frac{D\mathbf{q}^2}{\epsilon^2 + (D\mathbf{q}^2)^2}. \tag{8.14}$$

The correction to the density of states is thus negative. For short-range potentials, we can treat $U(\mathbf{q})$ as approximately constant provided the momentum is not too large. Since we are interested in the correction to density of states at low energies, the relevant momenta in Eq.(8.14) are less than $\sqrt{\epsilon/D}$, which, being small, allows us to treat $U(\mathbf{q})$ as a constant in this range. Doing a simple power counting, it is easily seen that in three dimensions, $\delta\nu(\epsilon)/\nu(0) \sim -\sqrt{|\epsilon|}$, while in two dimensions, $\delta\nu(\epsilon)/\nu(0) \sim \ln|\epsilon|$. Further details, such as the finite temperature behavior of the density of states, can be found in an excellent review by Altshuler and Aronov.[1]

8.2 Renormalization of the electron relaxation rate

In Sec. 8.1, we obtained the correction to the density of quasiparticle states through its dependence on the real part of the electron self energy Σ_ϵ. In the leading order in the perturbative expansion, the self energy is purely real. This does not affect the lifetime of the quasiparticle. We have seen earlier that in a clean Landau Fermi liquid, electron-electron collisions result in a finite lifetime $\tau_{\rm ee}$ for the quasiparticles. If the quasiparticle energy is ϵ, then $\tau_{\rm ee}^{-1} \propto \epsilon^2$. The vanishing of the quasiparticle decay rate faster than the quasiparticle energy underpins the validity of the Landau Fermi liquid theory. Since interparticle collisions are essential to relax the quasiparticle, the effect of electron-electron interactions on the lifetime appears at higher orders in the perturbation expansion. We will see below that even in the presence of weak potential disorder, the electron relaxation rate is much larger than the result for non-interacting electrons, and could even undermine the Fermi liquid description in low spatial dimensions.

Using the Fermi Golden rule for the decay rate due to electron-electron interaction followed by disorder averaging[2] (similar to Sec. 8.1), or using the Boltzmann transport approach for the distribution function of the diffusive electrons[3], it has been shown that

$$\frac{\partial n(\epsilon)}{\partial t} = \nu(0) \int_{-\infty}^{\infty} \frac{d\epsilon'}{\pi} \int_{-\infty}^{\infty} d\omega \int (d\mathbf{q}) |e^2 U(\mathbf{q},\omega)|^2 \left[\mathrm{Re} \frac{1}{D\mathbf{q}^2 - i\omega} \right]^2$$
$$\times \left[n_\epsilon n_{\epsilon'} (1 - n_{\epsilon-\omega})(1 - n_{\epsilon'+\omega}) - n_{\epsilon-\omega} n_{\epsilon'+\omega} (1 - n_\epsilon)(1 - n_{\epsilon'}) \right], \quad (8.15)$$

Here ϵ and ϵ' are quasiparticle energies of the incoming particles and ω is the energy transferred during scattering. The dominant contribution in momentum space comes from small momentum transfers, $q \sim \sqrt{\omega/D} \equiv L_\epsilon^{-1}$, where we have introduced the diffusion length L_ϵ. In this regime, we may approximate $U(\mathbf{q},\omega) \approx -1/\Pi(\mathbf{q},\omega) \sim 1/(e^2\nu(0))$.

[1]B. L. Altshuler and A. G. Aronov, (1985). Electron-electron interactions in disordered conductors, in *Electron-electron Interactions in Disordered Systems*, edited by A. L. Efros and M. Pollak (North Holland, Amsterdam), p. 1153.

[2]E. Abrahams, P. W. Anderson, P. A. Lee and T. V. Ramakrishnan, *Phys. Rev.* B **24**, 6783 (1981).

[3]A. Schmid, *Z. Phys.* **271**, 251 (1974).

At $T = 0$, Eq.(8.15) simplifies further. In particular, if we consider the $(\epsilon, \epsilon') \to (\epsilon - \omega, \epsilon' + \omega)$ process, the integration limits become $0 \le \omega \le \epsilon$ and $-\omega \le \epsilon' \le 0$. This leads us to the following expression for the decay rate due to electron-electron collisions,

$$\frac{1}{\tau_{\text{ee}}(\epsilon)} = \nu(0) \int_0^{\epsilon} \frac{d\omega}{2\pi} \int_{-\omega}^{0} d\epsilon' \int (d\mathbf{q}) |U(\mathbf{q}, \omega)|^2 \left[\text{Re} \frac{1}{D\mathbf{q}^2 - i\omega} \right]^2 . \qquad (8.16)$$

Employing the approximation $U(\mathbf{q}, \omega) \sim 1/\nu(0)$ in Eq.(8.16) and performing the integrals over energy and momentum, we finally arrive at the following estimate for the electron relaxation rate:

$$\frac{1}{\tau_{\text{ee}}(\epsilon)} \sim \frac{1}{\nu(0) L_{\epsilon}^{D}} . \qquad (8.17)$$

Here D is the spatial dimensionality. Thus even for $D = 3$, the electron relaxation rate for the diffusive case, $\tau_{\text{ee}}^{-1} \sim \epsilon^{3/2}$, is greater than $\tau_{\text{ee}}^{-1} \sim \epsilon^2$ for the clean Fermi liquid. Quasiparticles are however still well-defined. For $D = 2$, the electron relaxation rate for the diffusive case is comparable to the quasiparticle energy, and Fermi liquid theory is only marginally valid. At finite temperatures, for $D = 2$, the electron relaxation rate has a logarithmic divergence[4], $\tau_{\text{ee}}^{-1} \sim (k_B T) \ln |1/T|$.

In this chapter, our study of the electron relaxation rate has rested on the validity of the Fermi Golden Rule. Instead of our infinite system, let us now consider a large but finite system such that the mean level spacing is small but finite. The Golden Rule approximation is justified provided the density of final states is large enough such that the quasiparticle decay rate (defined through the width of the quasiparticle spectral peaks) exceeds the mean level spacing. In mesoscopic systems where the level separation becomes large enough to negate the Golden Rule approximation, the question of quasiparticle decay rate must once again be revisited[5]. We will discuss this problem later in Chapter 12.

[4]E. Abrahams, P. W. Anderson, P. A. Lee and T. V. Ramakrishnan, *Phys. Rev.* B **24**, 6783 (1981).

[5]See for example B. L. Altshuler, Y. Gefen, A. Kamenev and L. S. Levitov, *Phys. Rev. Lett.* **78**, 2803 (1997)

Chapter 9

Bridging Phenomenological and Microscopic Aspects

In this chapter, we provide a microscopic basis to the phenomenological understanding of Landau Fermi liquids that we developed earlier. A key assumption is that we can use perturbation theory to infinite order to treat the interaction of the particles.

This chapter is intended for the more advanced reader, who has some familiarity with the methods of quantum field theory, in particular the Feynman diagram technique. The advantage of the diagram technique is that it allows us to represent the summation of an arbitrary number of terms in the perturbation theory as a kind of graphical summation of diagrams. These diagrams, representing a sum of different terms in the perturbation theory, in turn appear as building blocks for constructing higher order contributions[1].

[1]For the diagram calculations, we use the convention of A. A. Abrikosov, L. P. Gorkov and I. E. Dzyaloshinskii, *Methods of Quantum Field Theory in Statistical Physics*, 2nd ed., Pergamon Press (1965). There are numerous excellent texts on many-body techniques in physics, and any of those will also be adequate for our purposes.

9.1 Proper self-energy Σ and vertex function Γ

Our starting point is the non-interacting (or bare) electron Green function (in Fourier space)

$$G^{(0)}(p) = \frac{1}{\epsilon - \xi^{(0)}(\mathbf{p}) + i\delta \text{ Sgn } \xi}. \tag{9.1}$$

Here $p = (\epsilon, \mathbf{p})$ is the "four-momentum" consisting of the energy ϵ and momentum $\mathbf{p}$ components, and $\xi^{(0)}(\mathbf{p}) = \epsilon^{(0)}(\mathbf{p}) - \mu$ is the kinetic energy of the electron reckoned from the Fermi level. The bare Green function will be diagrammatically denoted by a thin solid line.

Particle-type excitations, with $\xi^{(0)}(\mathbf{p}) > 0$, appear as poles, $\epsilon = \xi^{(0)}(\mathbf{p}) - i\delta$, of the retarded Green function, $G_R^{(0)} = 1/(\epsilon - \xi^{(0)}(\mathbf{p}) + i\delta)$, in the lower half of the complex ϵ-plane, while hole-type excitations, with $\xi^{(0)}(\mathbf{p}) < 0$, appear as poles of the advanced Green function, $G_A^{(0)} = 1/(\epsilon - \xi^{(0)}(\mathbf{p}) - i\delta)$, in the upper half of the complex ϵ-plane.

Two-particle interactions, $U^{(0)}(p)$, mediated by bosons (e.g., phonons or photons), will be shown as "vertices" with two incoming and two outgoing four-momenta. The total four-momentum entering or exiting such vertices is zero, in line with conservation of energy and momentum.

The interactions change the single-particle electron energies. The full Green function for single electrons then assumes the form

$$G(p) = \frac{1}{\epsilon - \xi^{(0)}(\mathbf{p}) - \Sigma(p)} = \frac{1}{\epsilon - \xi^{(0)}(\mathbf{p}) - \text{Re } \Sigma(p) - i \text{ Im } \Sigma(p)} \tag{9.2}$$

$$\equiv ((G^{(0)})^{-1} - \Sigma)^{-1}, \tag{9.3}$$

where Σ is called the proper self-energy (henceforth simply referred to as the self-energy). Equation (9.3) is an example of a Dyson equation. It relates the full Green function to the proper self energy. We will denote the full Green function by a thick solid line. Poles of $G(p)$ give the energies of the single-particle excitations in the interacting problem. As for the non-interacting case, the particle type excitations appear as poles of the retarded Green function G_R in the lower half of the ϵ-plane, below the $\epsilon < 0$ semi-axis. Similarly hole type excitations appear as poles of the advanced Green function G_A in the upper half of the ϵ-plane, above the $\epsilon > 0$ semi-axis. We have already seen that at least in three dimensions, for short-range interactions, the quasiparticle relaxation rate for an excitation of energy ϵ, $1/\tau_{\text{ee}}(\epsilon) \propto \text{Im } \Sigma_\epsilon$, vanishes as ϵ^2 in the absence of disorder and as $|\epsilon|^{3/2}$ for the diffusive case. This is smaller than the quasiparticle energy ϵ, which means that even in the presence of interactions, the single-particle excitations are well-defined. Combining with the fact that G_R and G_A are analytic continuations of G around different semi-axes of the ϵ plane, we can say that near $\epsilon = 0$,

$$G(p) \approx \frac{\mathcal{Z}}{\epsilon - v(|\mathbf{p}| - p_F) + i\delta \text{ Sgn } (|\mathbf{p}| - p_F)}. \tag{9.4}$$

The quantities $\mathcal{Z}$ and v in Eq.(9.4) are related to $\xi^{(0)}(\mathbf{p})$ and $\Sigma(p)$ in Eq.(9.2). Consider Eq.(9.2). Close to $\epsilon = 0$, (i.e., for momenta close to p_F), the pole of the Green function, $\epsilon = \xi(\mathbf{p})$, occurs at

$$\xi(\mathbf{p}) \approx \xi^{(0)}(\mathbf{p}) + \text{Re } \Sigma(\xi(\mathbf{p}), \mathbf{p}). \tag{9.5}$$

We ignored the imaginary part of the self-energy in Eq.(9.5) for at low energies that is much smaller than the real part. For momentum conserving interactions, while the Fermi momentum remains the same as the noninteracting case (we will provide a microscopic proof of this claim in Sec. 9.4), the Fermi energy μ does get shifted; its new position is given by demanding that quasiparticle energies vanish at the Fermi surface:

$$\xi(p_F) = \xi^{(0)}(p_F) + \text{Re } \Sigma(0, p_F) = 0. \tag{9.6}$$

Next we expand the self-energy in the vicinity of $\epsilon = \xi(\mathbf{p})$. The real part is

$$\text{Re } \Sigma(\epsilon, \mathbf{p}) \approx \text{Re } \Sigma(\xi(\mathbf{p}), \mathbf{p}) + (\epsilon - \xi(\mathbf{p}))(\partial \text{Re } \Sigma(\epsilon, \mathbf{p})/\partial\epsilon)|_{\epsilon=\xi(\mathbf{p})}. \tag{9.7}$$

Thus we may write $\epsilon - \xi^{(0)}(\mathbf{p}) - \text{Re } \Sigma(\epsilon, \mathbf{p}) \approx (\epsilon - \xi(\mathbf{p}))/\mathcal{Z}$,

$$\mathcal{Z}^{-1} \approx (1 - \partial \text{Re } \Sigma(\epsilon, \mathbf{p})/\partial\epsilon)|_{\epsilon=\xi(\mathbf{p})}. \tag{9.8}$$

We have dropped the explicit momentum dependence of $\mathcal{Z}$, as it is understood the momenta lie close to p_F. In the literature, the quantity $\mathcal{Z}$ is referred to as the quasiparticle weight. Finally, to obtain the renormalized Fermi velocity, v, we consider first the group velocity, $\mathbf{v} = \nabla_{\mathbf{p}}\xi(\mathbf{p})$. Taking the momentum gradient in Eq.(9.5) we get

$$\mathbf{v} = \frac{\partial\xi^{(0)}(\mathbf{p})}{\partial\mathbf{p}} + \left[\frac{\partial \text{Re } \Sigma(\epsilon, \mathbf{p})}{\partial\mathbf{p}} + \frac{\partial \text{Re } \Sigma(\epsilon, \mathbf{p})}{\partial\epsilon}\frac{\partial\xi(\mathbf{p})}{\partial\mathbf{p}}\right]_{\epsilon=\xi(\mathbf{p})}. \tag{9.9}$$

After some rearranging, and using Eq. (9.8),

$$\mathbf{v} = \mathcal{Z}\left[\frac{\partial\xi^{(0)}(\mathbf{p})}{\partial\mathbf{p}} + \left.\frac{\partial \text{Re } \Sigma(\epsilon, \mathbf{p})}{\partial\mathbf{p}}\right|_{\epsilon=\xi(\mathbf{p})}\right]. \tag{9.10}$$

The Fermi velocity v is the magnitude of the group velocity evaluated at the Fermi level, $\xi(\mathbf{p}) = 0$. Eq.(9.10) also gives us the the effective mass since $m^* = p_F/v$.

Similar to the expansion for the real part of the self-energy, we approximate the imaginary part of the self-energy in the vicinity of $\epsilon = \xi(\mathbf{p})$:

$$\text{Im } \Sigma(p) \approx \text{Im } \Sigma(\xi(\mathbf{p}), \mathbf{p}) = -|\text{Im } \Sigma(\xi(\mathbf{p}), \mathbf{p})|\text{Sgn } \xi(\mathbf{p}). \tag{9.11}$$

Eqs. (9.7) to (9.11) establish the relation between Eqs. (9.2) and (9.4). In the second equation in Eq.(9.11), we made use of the analytic properties of $G(p)$, namely, that particle type excitations ($\xi(\mathbf{p}) > 0$) appear as poles of $G(p)$ in the lower half of the ϵ plane, while hole type excitations appear as poles in the upper half plane. As a corollary of the above analysis, we also obtain the relation between the electron relaxation rate, $1/\tau_{\text{ee}}$, and Im Σ :

$$\frac{1}{\tau_{\text{ee}}(\xi(\mathbf{p}))} = -2\mathcal{Z}\text{Im }\Sigma(\xi(\mathbf{p}), \mathbf{p}). \tag{9.12}$$

The self-energy Σ contains, in general, perturbative contributions up to arbitrary order in the interactions. An important fact is that only certain diagrams appear in the perturbation expansion of Σ. These are the so-called irreducible diagrams, i.e., those that cannot be broken into two, or disconnected, by severing a single $G^{(0)}$ line. To see this is consistent with the perturbation expansion for the full Green function, we expand G in increasing powers of the interaction. This will in general contain both irreducible diagrams (those that cannot be broken into two by cutting a single $G^{(0)}$ line) as well as reducible ones. We collect the irreducible diagrams in Σ, which allows us to write the perturbation series for G as $G = G^{(0)} + G^{(0)}\Sigma G^{(0)} + G^{(0)}\Sigma G^{(0)}\Sigma G^{(0)} + \cdots$. This is nothing but Eq. (9.2).

9.1.1 Vertex function

Perturbative corrections to the bare interaction vertices $\Gamma^{(0)}$ result in the full interaction vertices Γ. In the diagrams, we will represent the bare vertex functions by open squares and full ones by filled squares. In principle, the proper self-energy contains all the information we need for computing the physical properties of the interacting electron gas. However in practice, proper self-energies usually require summation of an infinite number of diagrams. In such a case, it is convenient to express Σ through its relation to the vertex functions Γ. The quantities Σ and Γ, when they appear as part of other diagrams, obey the standard rules of perturbation theory; in particular, a factor of i for every vertex and a factor of (-1) for each fermion loop. In the rest of the chapter, we will assume short-range interactions unless stated otherwise. With long-range interactions (e.g. unscreened Coulomb), special care needs to be taken for interaction lines corresponding to small momenta.

Denote by $\Gamma_{\alpha\beta,\gamma\delta}(p, p'; p + q, p' - q)$ the vertex function describing the interaction of two particles with incoming four-momenta p, p' and outgoing four-momenta $p + q$ and $p' - q$ respectively, such that the four-momentum transferred, $q = (\omega, \mathbf{q})$, is small compared to p and p'. The Greek indices denote spins of the particles and they are to be taken in the same order as the momenta. We are interested in small q, i.e., $\omega \ll \mu$ and $|\mathbf{q}| \ll p_F$. The first order corrections to $\Gamma(p, p'; q)$ are shown in Fig.9.1. As the momentum transfer $q \to 0$, the poles of the internal Green function lines in diagram (c) merge, and this, as we will find below, leads to singularities in the perturbative corrections to Γ. Diagrams (a) and (b) show no singular behavior as $q \to 0$. We collect all such non-singular contributions into a new vertex $I(p, p'; q)$, the irreducible vertex[2]. As a result, we may safely put $q = 0$ in $I(p, p'; q)$. The vertex Γ therefore satisfies[3,4]

$$\Gamma_{\alpha\beta,\gamma\delta}(p, p'; q) = I_{\alpha\beta,\gamma\delta}(p, p') - i \int (dq') I_{\alpha\xi,\gamma\eta}(p, q') G(q + q') G(q') \Gamma_{\eta\beta,\xi\delta}(q', p'; q). \tag{9.13}$$

We made use of the fact that the vertices follow the standard Feynman diagram rules. Equation (9.13) is the Bethe-Salpeter equation for the four-legged vertex function, and is another example of a Dyson equation (the first one we encountered was in Eq. (9.3). We will henceforth suppress the spin labels for convenience unless necessary.

Let us examine the structure and singularities of the two-particle Green function, $G(q + q')G(q')$ that appears in the Bethe-Salpeter equation, Eq.(9.13).

9.2 ω/k singularities of the vertex function, relation to Landau f-functions

For small values of the four-momentum $q = (\omega, \mathbf{q})$, the singular contribution in Eq.(9.13) comes from integration in the vicinity of the poles of $G(q + q')G(q')$. The remaining nonsingular part in the integral may be assumed to vary slowly with q. The poles contribute only if both of them do not lie on the same side of the real axis. This either requires $|\mathbf{q}'| < p_F$ and $|\mathbf{q}' + \mathbf{q}| > p_F$, or the other way around. Using Eq.(9.4) for the form of the Green function near its poles, the two-particle Green function is approximately given by

$$G(q + q')G(q') \approx \frac{\mathcal{Z}^2 \Theta(-\xi(\mathbf{q}'))\Theta(\xi(\mathbf{q} + \mathbf{q}'))}{(\epsilon - v(|\mathbf{q}'| - p_F) - i\delta)(\epsilon + \omega - v(|\mathbf{q}' + \mathbf{q}| - p_F) + i\delta)} + \frac{\mathcal{Z}^2 \Theta(\xi(\mathbf{q}'))\Theta(-\xi(\mathbf{q} + \mathbf{q}'))}{(\epsilon - v(|\mathbf{q}'| - p_F) + i\delta)(\epsilon + \omega - v(|\mathbf{q}' + \mathbf{q}| - p_F) - i\delta)} + R(q'). \tag{9.14}$$

[2]Thus no pair of internal Green function lines in the irreducible vertex differ by a small momentum.

[3]P. Nozières and J. M. Luttinger, *Phys. Rev.* **127**, 1423 (1962).

[4]A. A. Abrikosov, L. P. Gorkov and I. E. Dzyaloshinskii, *Methods of Quantum Field Theory in Statistical Physics*, 2nd ed., Pergamon Press (1965).

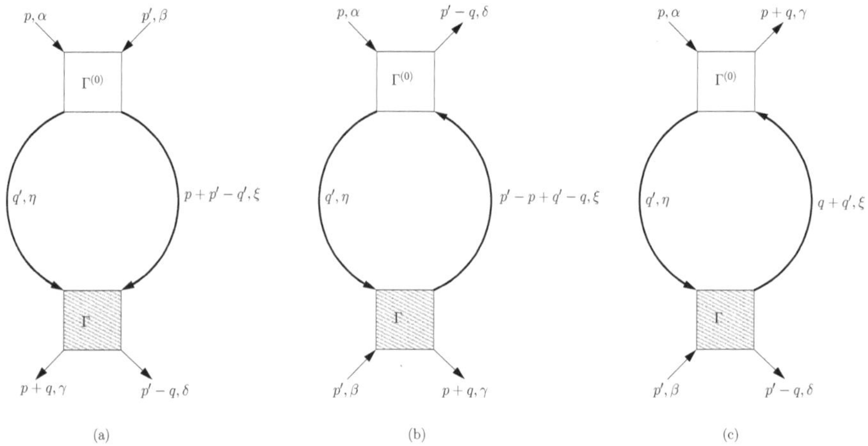

FIGURE 9.1: Diagrams contributing to leading perturbative corrections to the vertex function $\Gamma_{\alpha\beta,\gamma\delta}(p, p'; p+q, p'-q)$ (shown as full squares) for small values of the four-momentum q. The open squares represent the bare vertex $\Gamma^{(0)}$. The external legs indicate the incoming and outgoing four-momenta p, p', q and spins (Greek indices) at the vertices. Thick solid lines represent the full Green function G. In (c), the poles of the internal Green functions merge as $q \to 0$, which leads to singularities in Γ. Diagrams (a) and (b) exhibit no singularity as $q \to 0$. Internal four-momenta q' and spins η, ξ are to be summed over. The irreducible vertex $I(p, p')$ introduced in the text contains, besides $\Gamma^{(0}$, all diagrams that do not contain pairs of internal Green function lines with close momenta.

In Eq.(9.14), the four-momentum $q' = (\epsilon, \mathbf{q}')$, with the poles near $\epsilon \approx 0$ and $|\mathbf{q}'| \approx p_F$. The quantity $R(q')$ contains the non-singular contributions in $G(q+q')G(q')$ as $q \to 0$. The We now use the identities $1/(x \pm i\delta) = P(1/x) \mp i\pi\delta(x)$ and obtain the contribution near the poles in Eq.(9.14):

$$G(q+q')G(q') \approx i2\pi\mathcal{Z}^2\delta(\epsilon)\left[\frac{\Theta(\xi(\mathbf{q}+\mathbf{q}')) - \Theta(\xi(\mathbf{q}'))}{\omega + v(|\mathbf{q}'| - |\mathbf{q}'+\mathbf{q}|)}\right] + R(q') \tag{9.15}$$

$$= i2\pi\mathcal{Z}^2\delta(\epsilon)\delta(\xi(\mathbf{q}'))\frac{\mathbf{v}_{\mathbf{q}'}\cdot\mathbf{q}}{\omega - \mathbf{v}_{\mathbf{q}'}\cdot\mathbf{q}} + R(q'). \tag{9.16}$$

The RHS of Eq.(9.15) is closely related to the Lindhard polarizability, $\Pi(\omega, \mathbf{q})$, which follows from a summation over the internal four-momentum q' (see Eq. (7.2)).

A crucial property of the two-particle Green function evident from Eq.(9.16) is that it essentially depends on the order in which the $\omega \to 0$ and $\mathbf{q} \to 0$ limits are taken:

$$G(q+q')G(q')|_{q\to 0} = \begin{cases} R(q'), & |\mathbf{q}|/\omega \to 0 \\ R(q') - i2\pi\mathcal{Z}^2\delta(\epsilon)\delta(\xi(\mathbf{q}')), & \omega/|\mathbf{q}| \to 0. \end{cases} \tag{9.17}$$

Let us examine the Bethe-Salpeter equation, Eq.(9.13) in these two limits. Consider first, using Landau's parlance, the ω-limit ($\omega \to 0$ and $|\mathbf{q}|/\omega \to 0$). Denoting the vertex function in this limit by Γ^ω,

$$\Gamma^\omega(p, p'; q) = I(p, p') - i\int (dq')I(p, q')R(q')\Gamma^\omega(q', p'; q). \tag{9.18}$$

Symbolically we write the Bethe-Salpeter equations for Γ^ω and Γ as

$$\begin{aligned} \Gamma^\omega &= I - iIR\Gamma^\omega, \\ \Gamma &= I - iI(R + i\Phi)\Gamma, \end{aligned} \tag{9.19}$$

where

$$\Phi(q, q') = 2\pi\mathcal{Z}^2\delta(\epsilon)\delta(\xi(\mathbf{q}'))\frac{\mathbf{v}_{\mathbf{q}'}\cdot\mathbf{q}}{\omega - \mathbf{v}_{\mathbf{q}'}\cdot\mathbf{q}} \tag{9.20}$$

is the (singular) imaginary part of the two-particle Green function. We use Eqs. (9.19) to eliminate the irreducible vertex I and the non-singular part R. From the first equation we have $\Gamma^\omega = (1 + iIR)^{-1}I$, and from the second, $(1 + iIR)\Gamma = I(1 + \Phi\Gamma)$. Consequently, the Bethe-Salpeter equation becomes

$$\Gamma = \Gamma^\omega + \Gamma^\omega\Phi\Gamma, \text{ or,} \tag{9.21}$$

$$\Gamma = (1 - \Gamma^\omega\Phi)^{-1}\Gamma^\omega. \tag{9.22}$$

For the spin-independent case, the quantity $(1-\Gamma^\omega\Phi)$ has the physical meaning of the dielectric function ε. In general, Eq. (9.22) expresses screening of the vertex Γ^ω by low-energy particle-hole excitations (represented by Φ). The quantity Φ, as we have seen earlier, has the same form as the small momentum behavior of the Lindhard polarizability, Π, of the interacting electron system. In analogy with Eq. (7.2), collective excitations of the system are given by zeros of $(1 - \Gamma^\omega \Phi)$, or equivalently, the poles of the vertex function Γ.

Next, consider the **q**-limit ($|\mathbf{q}| \to 0$ and $\omega/|\mathbf{q}| \to 0$) in Eq. (9.21). In the **q**-limit, $\Phi(q, q') \approx -2\pi \mathcal{Z}^2 \delta(\epsilon)\delta(\xi(\mathbf{q}'))$. Denoting the corresponding vertex function by $\Gamma^{\mathbf{q}}$, and using Eq.(9.20) for Φ, the Bethe-Salpeter equation assumes the form

$$\Gamma^{\mathbf{q}}(p, p'; q) = \Gamma^\omega(p, p') - 2\pi \mathcal{Z}^2 \int (dq')\Gamma^\omega(p, q')\delta(\epsilon)\delta(\xi(\mathbf{q}'))\Gamma^{\mathbf{q}}(q', p'; q). \tag{9.23}$$

Here $(dq') = d\epsilon\, d\Omega_{\mathbf{q}'}\, d|\mathbf{q}'||\mathbf{q}'|^2/(2\pi)^4$. The $\delta(\xi(\mathbf{q}')) = (1/v)\delta(|\mathbf{q}'| - p_F)$ function constrains the momentum integration to the Fermi surface; thus,

$$\Gamma^{\mathbf{q}}(p, p'; q) = \Gamma^\omega(p, p') - \frac{\mathcal{Z}^2 p_F^2}{v(2\pi)^3} \int d\Omega_{\mathbf{q}'} \Gamma^\omega(p, (0, p_F \hat{\mathbf{q}}'))\Gamma^{\mathbf{q}}((0, p_F \hat{\mathbf{q}}'), p'; q). \tag{9.24}$$

Similarly performing the integration over the independent internal momenta in Eq.(9.21), the full vertex function Γ is found to obey the integral equation,

$$\Gamma(p, p'; q) = \Gamma^\omega(p, p') + \frac{\mathcal{Z}^2 p_F^2}{v(2\pi)^3} \int d\Omega_{\mathbf{q}'} \Gamma^\omega(p, q') \frac{\mathbf{v}_{\mathbf{q}'} \cdot \mathbf{q}}{\omega - \mathbf{v}_{\mathbf{q}'} \cdot \mathbf{q}} \Gamma(q', p'; q). \tag{9.25}$$

Equations (9.24) and (9.25) are key equations needed for a microscopic theory of Landau Fermi liquids.

To solve Eq.(9.25) for $\Gamma(p, p'; q)$, we first note that for small values of q, the RHS is dominated by the second term because of the presence of the singular part of the two-particle Green function. We therefore discard the first term on the RHS, Γ^ω. We then try the ansatz

$$\Gamma(p, p'; q) = \chi(p, q)\chi(p', q) \tag{9.26}$$

in Eq. (9.25) which gives us

$$\chi(p, q) = \frac{\mathcal{Z}^2 p_F^2}{v(2\pi)^3} \int d\Omega_{\mathbf{q}'} \Gamma^\omega(p, q') \frac{\mathbf{v}_{\mathbf{q}'} \cdot \mathbf{q}}{\omega - \mathbf{v}_{\mathbf{q}'} \cdot \mathbf{q}} \chi(q', q). \tag{9.27}$$

Further simplification is possible by introducing the quantity $\hat{\nu}$:

$$\hat{\nu}(\hat{\mathbf{q}}') = \frac{\hat{\mathbf{q}}' \cdot \mathbf{q}}{\omega - v\hat{\mathbf{q}}' \cdot \mathbf{q}} \chi(q', q). \tag{9.28}$$

Substitution of $\hat{\nu}$ into Eq.(9.27) leads us to the following integral equation for $\hat{\nu}$:

$$(\omega - v\hat{\mathbf{p}} \cdot \mathbf{q})\hat{\nu}(\hat{\mathbf{p}}) = (\hat{\mathbf{p}} \cdot \mathbf{q}) \frac{\mathcal{Z}^2 p_F^2}{(2\pi)^3} \int d\Omega_{\mathbf{q}'} \Gamma^\omega(\mathbf{p}, \mathbf{q}')\hat{\nu}(\hat{\mathbf{q}}'). \tag{9.29}$$

Equation (9.29) is essentially the same as Eq. (3.12) that we obtained in our phenomenological study of zero sound in Landau Fermi liquids. We also identify the microscopic quantity $\mathcal{Z}^2\Gamma^\omega(\mathbf{p},\mathbf{q}')$ as the Landau f-function.

It turns out that the vertex function Γ is related to the scattering amplitude of two quasiparticles. To see this, recall from Chapter 4 that the scattering amplitude is proportional to the T-matrix, which in the first Born approximation is simply the bare interaction[5] subject to the constraint that the time-like components of the four-momenta (i.e. frequencies) associated with the incoming and outgoing legs are equal to the respective energies $\xi(\mathbf{p}_i)$. Higher order contributions to T (using, say, the Lippman-Schwinger formalism we discussed in Chapter 4) generate diagrams identical to those appearing in the perturbation theory for Γ. In the Fermi liquid, the incoming and outgoing states are not quite "noninteracting", a fact that shows up in the quasiparticle weight $\mathcal{Z}$. The T-matrix (or scattering amplitude) is thus

$$T(\mathbf{k}_1,\mathbf{k}_2;\mathbf{k}_3,\mathbf{k}_4) = \mathcal{Z}^2\Gamma(k_1,k_2;k_3,k_4)|_{\epsilon_i=\xi(\mathbf{p}_i)}. \tag{9.30}$$

The vertex function in the $\mathbf{q}$-limit, $\Gamma^{\mathbf{q}}$ evaluated at the Fermi surface (i.e. $\epsilon_i = \xi(\mathbf{p}_i) = 0$) is now interpreted as the forward scattering amplitude for particles at the Fermi surface, $T(\mathbf{p},\mathbf{p}') = \mathcal{Z}^2\Gamma^{\mathbf{q}}(\mathbf{p},\mathbf{p}')$. It follows that the Landau f-function, that we argued above as related to the vertex function in the ω-limit through $f \sim \mathcal{Z}^2\Gamma^\omega$, can be expressed in terms of the scattering amplitude through Eq.(9.24).

9.3 Ward identities

Ward identities are exact relations between correlation or response functions and are the result of symmetries or conservation laws. In Eq. (9.12) we obtained the quasiparticle relaxation rate in terms of the imaginary part of the electron self-energy Σ. The renormalized Fermi velocity and quasiparticle effective mass were also expressed in terms of the electron self-energy (see Eq.(9.10)). On the other hand, the phenomenology of the Landau Fermi liquids is built around interaction parameters – the Landau f-functions, which in Sec.9.2 we related to the irreducible vertex function in the ω-limit, Γ^ω. Ward identities provide relations between the self-energy and the vertex functions, thus bridging the phenomenological and microscopic aspects. Ward identities are also crucial in understanding the electromagnetic response of Fermi liquids. A key assumption in the Landau Fermi liquid theory is that the quasiparticles couple to external electromagnetic fields with the same strength as

[5] The bare interaction is also the leading order term in the perturbation expansion for the vertex function.

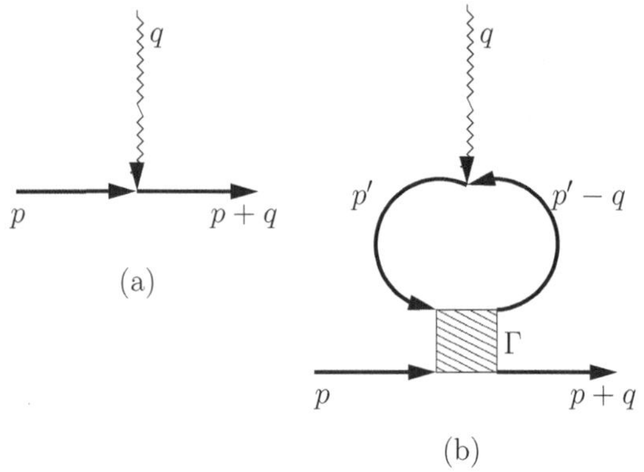

FIGURE 9.2: Diagrams for the correction to the Green function, ΔG, due to an infinitesimal external scalar potential $\Delta V(q)$ carrying an "inward" four-momentum q. The thick lines denote the full Green functions and the shaded square is the vertex function, Γ. The diagrams (a) and (b), combined with gauge invariance, are used to obtain the Ward identity of Eq. (9.33).

the noninteracting electrons. The microscopic justification of this assumption rests on Ward identities.

The Ward identities we will discuss readily follow from the definition of the vertex function Γ. For simplicity, we will discuss spinless electrons in this section. Figure 9.2 shows diagrams for the change in the Green function, ΔG, on account of introducing an infinitesimal external scalar potential $\Delta V(q)$ carrying an "inward" four-momentum $q = (\omega, 0)$ The diagrams for ΔG consist of an incoming and an outgoing leg, carrying four-momenta $p = (\epsilon, \mathbf{p})$ and $p+q$ respectively. From Fig. 9.2, the change in the Green function is[6]

$$\Delta G = G(p)G(p+q)\Delta V(q) - iG(p)G(p+q)\Delta V(q)\int (dp')G(p')G(p'-q)\Gamma(p,p';q). \tag{9.31}$$

We are interested in the limit $\omega \to 0$. Because of gauge invariance, the change ΔG induced by a constant ΔV must equal the change ΔG in response to the shift $\epsilon \to \epsilon - \Delta V$. Using $\Delta G/\Delta V = -\partial G/\partial \epsilon$ in Eq. (9.31) we obtain for $\omega \to 0$,

$$\frac{\partial G(p)}{\partial \epsilon} = -\{G(p)^2\}_\omega + i\{G(p)^2\}_\omega \int (dp')\{G(p')^2\}_\omega \Gamma^\omega(p,p'). \tag{9.32}$$

Here $\{G(p)^2\}_\omega = G(p)G(p+(\omega,0))|_{\omega\to 0} \equiv R(p)$ is the non-singular part of

[6]Putting back the spin labels would give us $\Delta G_{\alpha\beta} = \delta_{\alpha\beta}G(p)G(p+q)\Delta V(q) - iG(p)G(p+q)\Delta V(q)\int(dp')G(p')G(p'-q)\Gamma_{\alpha\eta,\beta\eta}(p,p';q)$. Since $\Delta V(q)$ does not affect the spins, we can further simplify by taking the trace over the external spin label, which gives $\Delta G = G(p)G(p+q)\Delta V(q) - \frac{i}{2}G(p)G(p+q)\Delta V(q)\int(dp')G(p')G(p'-q)\Gamma_{\alpha\eta,\alpha\eta}(p,p';q)$.

$G(p)G(p+q)|_{q\to 0}$ (see Eq. (9.17)). Equation (9.32) may be tidied up a bit more by dividing both sides by $\{G(p)^2\}_\omega$, and making use of the relations $\partial G(p)^{-1}/\partial\epsilon = -\{G(p)^2\}_\omega \partial G(p)/\partial\epsilon$ and the Dyson equation $G^{-1} = G_0^{-1} - \Sigma$. We are thus led to our first Ward identity[7],

$$\frac{\partial \Sigma(p)}{\partial \epsilon} = i\int (dp')R(p')\Gamma^\omega(p,p'), \tag{9.33}$$

essentially a consequence of gauge invariance. Near the pole of $G(p)$, we have seen earlier that the real part of the self-energy is generally much larger than its imaginary part, and the latter vanishes at the Fermi surface. Since $\partial \text{Re}\ \Sigma/\partial\epsilon$ is related to $\mathcal{Z}$ through $\mathcal{Z}^{-1} = 1 - \partial\text{Re}\ \Sigma/\partial\epsilon|_{\epsilon=\xi(\mathbf{p})}$ (see Eq. (9.8)), we find that Eq. (9.33) also provides a relation between the vertex function in the ω-limit, Γ^ω, and $\mathcal{Z}$. Since the quasiparticle weight is determined by the non-singular (or non-quasiparticle) part of $G(p)G(p+q)|_{q\to 0}$, it may be viewed as a measure of the overlap of a single particle state and a quasiparticle state, with the latter involving one or more particle-hole pairs over and above the single particle excitation.

Let us now introduce an infinitesimal static vector potential, $\Delta\mathbf{A}(q)$, ($q = (0,\mathbf{q})$). Gauge invariance requires the replacement $\mathbf{p}\to\mathbf{p} - e\Delta\mathbf{A}(\mathbf{r})$, and therefore $\Delta G/e\Delta\mathbf{A} = -\partial G/\partial\mathbf{p}$. Arguments similar to those that led us to Eq. (9.33) give us our second Ward identity:

$$\frac{\partial \Sigma(p)}{\partial \mathbf{p}} = -i\int (dp')\{G(p')^2\}_\mathbf{q}\frac{\partial \xi^{(0)}(\mathbf{p}')}{\partial \mathbf{p}'}\Gamma^\mathbf{q}(p,p'). \tag{9.34}$$

In Eq. (9.10) we obtained an expression for the effective quasiparticle velocity, $\mathbf{v}$ (or equivalently the inverse of the effective mass, m^*), in terms of $\mathcal{Z}$ and the derivative of the real part of the self-energy with respect to the momentum. Equation (9.34) thus also enables us to obtain the relation between $|\mathbf{v}| = p_F/m^*$, the vertex function Γ^ω, and $\mathcal{Z}$. To obtain this relation, we eliminate $\{G(p)^2\}_\mathbf{q}$ in Eq. (9.34) using the identity

$$\{G(p)^2\}_\mathbf{q} = \{G(p)^2\}_\omega - \frac{i2\pi\mathcal{Z}^2}{v}\delta(\epsilon)\delta(|\mathbf{p}|-p_F), \tag{9.35}$$

and Eq. (9.24) that expresses $\Gamma^\mathbf{q}$ in terms of Γ^ω. It is straightforward to show

$$\frac{1}{m} = \frac{1}{m^*} + \frac{\mathcal{Z}^2 p_F}{2\pi^2}\int\frac{d\Omega}{4\pi}\cos\theta\Gamma^\omega(\theta), \tag{9.36}$$

[7] For the spinful case, our first Ward identity would become $\frac{\partial\Sigma(p)}{\partial\epsilon} = \frac{i}{2}\int(dp')R(p')\Gamma^\omega_{\alpha\eta,\alpha\eta}(p,p')$. To obtain this, we attach spin indices to the self-energy and vertices in Eq. (9.33), $\Sigma\to\Sigma_{\alpha\eta}$, $\Gamma\to\Gamma_{\alpha,\xi,\eta\xi}$, and take the trace over the spin indices of the external legs. The internal label ξ is summed over since the scalar potential couples to both spin species in the same manner.

where θ is the angle between $\mathbf{p}$ and $\mathbf{p}'$. Note that this result is for spinless electrons [8].

The third Ward identity concerns Galilean invariance. Consider another reference frame moving (with respect to the original frame) with a non-stationary velocity $\Delta\mathbf{u}$, which in Fourier space has the components $\Delta\mathbf{u}(q)$, $q = (\omega, 0)$. The energy of a quasiparticle with momentum $\mathbf{p}$ is shifted by $-\Delta\mathbf{u}\cdot\mathbf{p}$, which can be absorbed by shifting the frequency $\epsilon \to \epsilon + \Delta\mathbf{u}\cdot\mathbf{p}$. In the limit of a constant relative velocity, the change in the Green function may be expressed as $\Delta G = (\partial G(p)/\partial\epsilon)(\mathbf{p}\cdot\Delta\mathbf{u})$. Following similar arguments as those that resulted in the previous two Ward identities, Eqs. (9.33) and (9.34), we obtain our third Ward identity:

$$\mathbf{p}\frac{\partial\Sigma(p)}{\partial\epsilon} = i\int(dp')R(p')\mathbf{p}'\Gamma^{\omega}(p,p'). \tag{9.37}$$

A fourth Ward identity also appears occasionally in the literature,[9] where a static scalar potential $\Delta V(\mathbf{r})$ is introduced instead of the spatially uniform but time-dependent ΔV we considered in our derivation of the first Ward identity, Eq. (9.33). This case corresponds to the $\mathbf{q}$-limit (Eq. (9.33) pertains to the ω-limit). This fourth Ward identity is

$$\frac{\partial\Sigma(p)}{\partial\mu} = i\int(dp')\{G(p')^2\}_{\mathbf{q}}\Gamma^{\mathbf{q}}(p,p'). \tag{9.38}$$

As a corollary of the Ward identities, we are also able to provide a microscopic justification for ignoring the quasiparticle weight $\mathcal{Z}$ in electromagnetic response functions in the phenomenology of Landau Fermi liquids. Consider, for example, the vertex corresponding to the coupling of electrons to a small spatially uniform external scalar potential $\Delta V(q)$, $q = (\omega, 0) \to 0$. In Landau theory, the elementary excitations are the quasiparticles, and the electromagnetic field couples to these quasiparticle fields, and not the full electron field. One therefore expects the electromagnetic vertex to be dressed by $\mathcal{Z}$, which is not present in Landau theory. Now we include vertex corrections (i.e., Fig. 9.2(b) with external legs connected to quasiparticle fields). With the vertex corrections, the overall "dressing" of the electromagnetic vertex becomes $\mathcal{Z}[1 - i\int(dp')R(p')\Gamma^{\omega}(p,p')]$. From the discussion following Eq. (9.33), it is clear that the quantity $[1 - i\int(dp')R(p')\Gamma^{\omega}(p,p')]$ is nothing but $1/\mathcal{Z}$. This cancels the factor $\mathcal{Z}$ in the numerator, in agreement with Landau theory. Thus vertex corrections, through their relation to the Ward identity for the self-energy, enable us to provide a microscopic justification of one of the key assumptions of Landau Fermi liquid theory.

---

[8] For spinful electrons, we will get $\frac{1}{m} = \frac{1}{m^*} + \frac{\mathcal{Z}^2 p_F}{4\pi^2}\int\frac{d\Omega}{4\pi}\cos\theta\Gamma^{\omega}_{\alpha\beta,\alpha\beta}(\theta)$. Comparing with Eq. (1.43), one identifies $\mathcal{Z}^2\Gamma^{\omega}_{\alpha\beta,\alpha\beta}(\mathbf{p},\mathbf{p}') = \mathrm{Tr}_{\sigma}\mathrm{Tr}_{\sigma'}f(\mathbf{p},\boldsymbol{\sigma};\mathbf{p}',\boldsymbol{\sigma}')$.

[9] See for example A. A. Abrikosov, L. P. Gorkov and I. E. Dzyaloshinskii, *Methods of Quantum Field Theory in Statistical Physics*, 2nd ed., Pergamon Press (1965).

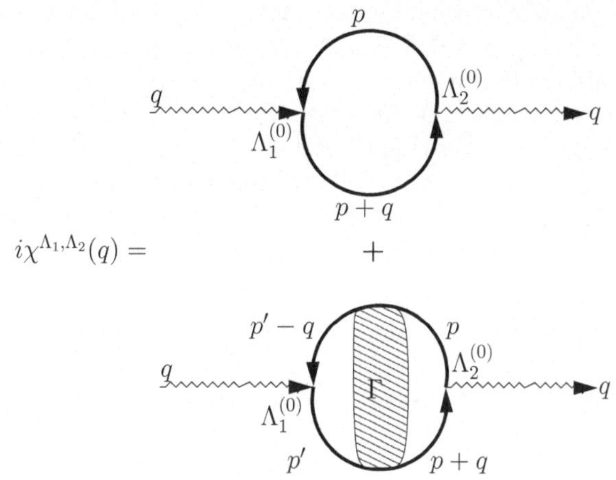

FIGURE 9.3: Perturbation expansion of the quasiparticle response function $\chi^{\Lambda_1,\Lambda_2}(q)$ in terms of the full quasiparticle Green functions, G, (solid lines) and the full vertex function Γ (shaded box). The $\Lambda_1^{(0)}(p',q)$ and $\Lambda_2^{(0)}(p,-q)$ are bare matrix elements associated with the coupling of the perturbation (wavy lines) with the electrons (see text for more details). These matrix elements depend on the four-momenta and spins of the scattering particles.

9.3.1 Response functions

We observed above in Eq. (9.15) that the singular part of the quantity $\int(dq')G(q+q')G(q')$, that is $i\int(dq')\Phi(q,q')$, is essentially $i(\mathcal{Z}^2/e^2)\times$ the bare Lindhard polarizability function ($\Pi(q)$) that determines the linear response of an interacting electron gas to an external electric potential (see Eq. (6.17)). The factor e^2 is the strength of the Coulomb interaction. In Fig. 9.2, we generalize the three-legged interaction vertex $V(q)$ to $\Lambda^{(0)}_{\sigma\sigma'}(p,q)V(q)$, where $\Lambda^{(0)}$ is the (bare) matrix element of the coupling and in general depends on the four-momenta of the scattering electrons and their spins. For example, it is $e\delta_{\sigma\sigma'}$ for the Coulomb interaction, $e\mathbf{p}\delta_{\sigma\sigma'}$ for coupling to a vector potential $\mathbf{A}(\mathbf{q})$, $\mu_B\boldsymbol{\sigma}_{\sigma\sigma'}$ for (spin) Zeeman coupling to a magnetic field, etc. The perturbation $\Lambda^{(0)}_{\sigma\sigma'}(p,q)V(q)$ couples to the density $\rho^{\Lambda}_{\sigma\sigma'}(p,q) = c^\dagger_\sigma(p+q)c_{\sigma'}(p)$ conjugate to it. The response function $\chi^{\Lambda_1,\Lambda_2}$ is defined as the conjugate density associated with Λ_1 induced in response to a perturbation described by $\Lambda_2^{(0)}$. The perturbation expansion of $\chi^{\Lambda_1,\Lambda_2}$ in their interaction is shown diagrammatically in Fig. 9.3. The right half of the diagrams in Fig. 9.3 are easily identifiable with the diagrams in Fig. 9.2 that we encountered earlier in connection with the Ward identity for the renormalization of the three-legged interaction vertex – including the vertex corrections in the ω-limit (i.e. spatially uniform and slowly varying in time) leads to the renormalization $\{\Lambda^{(0)}\}_\omega \to \Lambda^\omega = \Lambda^{(0)}/\mathcal{Z}$.

The two-particle internal Green functions in this limit reduce to the nonsingular part R in Eq. (9.15). We must remember, of course, that the Ward identity is applicable only if a conservation law is associated with perturbation corresponding to the relevant three-legged vertex. Thus for conserved spin currents the normalization factor is there but if spin is not conserved (as is often the case), then the corresponding Ward identity does not exist.

The diagrammatic expansion (or Bethe-Salpeter equation) for the response function $\chi^{\Lambda_1,\Lambda_2}$ is formally written as

$$\begin{aligned}\chi^{\Lambda_1,\Lambda_2}(q) = &-i\mathrm{Tr}_\sigma \int (dp)\Lambda_1^{(0)}(p,q)\Lambda_2^{(0)}(p,-q)G(p)G(p+q)\\ &- i\mathrm{Tr}_\sigma \int (dp)(dp')\Lambda_1^{(0)}(p',q)G(p')G(p'-q)\\ &\times -i\Gamma(p,p';q)\Lambda_2^{(0)}(p,-q)G(p)G(p+q),\end{aligned} \tag{9.39}$$

where Tr_σ denotes trace over the spin degrees of freedom. In the rest of this section, trace over spins is understood and will not be written explicitly. We may symbolically write Eq. (9.39) in terms of a two-particle correlator $\bar{\chi}$: $\chi = \Lambda_1^{(0)}\bar{\chi}\Lambda_2^{(0)}$, where

$$i\bar{\chi} = (R + i\Phi) + (R + i\Phi)(-i\Gamma)(R + i\Phi). \tag{9.40}$$

We introduce now the renormalized three-legged vertex functions, $\Lambda = (1 - i\Gamma(R + i\Phi))\Lambda^{(0)}$, and their ω-limit, $\Lambda^\omega = (1 - i\Gamma^\omega R)\Lambda^{(0)}$. Using these with the Bethe-Salpeter equation for the four-legged vertex, $\Gamma = \Gamma^\omega + \Gamma\Phi\Gamma^\omega$, we find that the renormalized three-legged vertex function also follows a Bethe-Salpeter equation, $\Lambda = \Lambda^\omega + \Gamma\Phi\Lambda^\omega$. The response function then takes the form

$$\begin{aligned}i\chi &= i\chi^\omega + \Lambda_1^\omega\Phi[1 + \Gamma\Phi]\Lambda_2^\omega\\ &= i\chi^\omega + \Lambda_1^\omega\Phi(1 - \Gamma^\omega\Phi)^{-1}\Lambda_2^\omega,\end{aligned} \tag{9.41}$$

and

$$i\chi^\omega = \int (dp)\left[R(p) - iR(p)\int (dp')\Gamma^\omega(p,p')R(p')\right]\Lambda_1^{(0)}(p')\Lambda_2^{(0)}(p). \tag{9.42}$$

In the second line of Eq. (9.41) we used the expression for Γ obtained in terms of Γ^ω in Eq. (9.22). Since Φ and Γ^ω both involve perturbative contributions to infinite order in the interaction, one needs to ensure that the expansion is consistent – this means the Γ^ω and quasiparticle weight $\mathcal{Z}$ must satisfy the appropriate Ward identities that enforce the conservation laws.

For the special cases in which one or both three-legged vertices in the response function are associated with a conserved charge, Ward identities can be used to further simplify the result in Eq. (9.41). For example, for the charge susceptibility, we have $\Lambda_1^{(0)} = \Lambda_2^{(0)} = e$. Consider in Eq. (9.41) the χ^ω term which is the contribution from the non-singular part (also known as incoherent part in the literature). From the Ward identity in Eq. (9.32), the quantity in square brackets is simply $-\partial G(p)/\partial\epsilon$, where $p = (\epsilon, \mathbf{p})$. Hence

$$i\chi^\omega = -e^2 \int (d\mathbf{p})(d\epsilon) \frac{\partial G(p)}{\partial \epsilon} \equiv 0. \tag{9.43}$$

We used the fact that the Green function vanishes as $\sim 1/\epsilon$ as $\epsilon \to \pm\infty$. Similarly in the context of the Ward identity associated with Galilean invariance, Eq. (9.37), we consider $\Lambda_1^{(0)}(p) = \Lambda_2^{(0)}(p) = \mathbf{p}$. Once again, using this particular Ward identity, it is easy to see that the corresponding χ^ω vanishes. In the same manner, it is also possible to verify the vanishing of χ^ω whenever $\Lambda_1^{(0)}$ or $\Lambda_2^{(0)}$ is associated with a conserved charge. Thus, in general, using the Ward identities involving the ω-limit, one finds that the susceptibility associated with conserved charges has a simple form,

$$i\chi^{\Lambda_1,\Lambda_2} = \Lambda_1^\omega \Phi (1 - \Gamma^\omega \Phi)^{-1} \Lambda_2^\omega. \tag{9.44}$$

9.4 Luttinger's theorem

Since momentum conserving interparticle interactions do not change the Fermi momentum, p_F, we expect that the number density of particles $n = N/V$, being proportional to the momentum-space volume of the Fermi sphere, is also unaffected by such interactions. This relation between the number density and the volume enclosed by the Fermi surface in momentum space is known as Luttinger's theorem.[10]

The number density is

$$\begin{aligned} n = \frac{N}{V} &= \langle \psi^\dagger(\mathbf{r},t)(\mathbf{r},t)\rangle \\ &= -i\left[-i\langle T(\psi(\mathbf{r},t)\psi^\dagger(\mathbf{r},t+\eta))\rangle|_{\eta\to 0^+}\right]. \end{aligned} \tag{9.45}$$

The quantity in square brackets is the usual time ordered Green function. Going over to momentum space,

$$\begin{aligned} n &= -i\int \frac{d\epsilon\, d^d\mathbf{p}}{(2\pi)^{d+1}}\, e^{i\epsilon\eta} G(\epsilon,\mathbf{p}) \\ &= i\int \frac{d\epsilon\, d^d\mathbf{p}}{(2\pi)^{d+1}}\, e^{i\epsilon\eta}\left[\frac{\partial}{\partial\epsilon}\ln G(p) - G(p)\frac{\partial}{\partial\epsilon}\Sigma(p)\right]. \end{aligned} \tag{9.46}$$

[10] J. M. Luttinger and J. C. Ward, *Phys. Rev.* **118**, 1417 (1960).

The next step is to show that the second contribution to the integral in Eq. (9.46), $\int_p e^{i\epsilon\eta}G(p)(\partial\Sigma(p)/\partial\epsilon)$, vanishes. Integrating by parts and using $G(\epsilon,\mathbf{p})|_{|\epsilon|\to\infty} = 1/\epsilon$, and $|\Sigma(\epsilon,\mathbf{p})|_{|\epsilon|\to\infty}$ remaining smaller than $|\epsilon|$, we have

$$\int (d\epsilon)G(p)\frac{\partial}{\partial\epsilon}\Sigma(p) = -\int (d\epsilon)\Sigma(p)\frac{\partial}{\partial\epsilon}G(p). \tag{9.47}$$

To proceed further, we make use of another fact, that the (proper) self-energies $\Sigma(p)$ may be expressed as a functional derivative with respect to $G(p)$ of the Luttinger-Ward functional, $\Phi[G]$. To construct $\Phi[G]$ satisfying this property, we consider proper self-energy diagrams consisting of irreducible vertices and full Green functions G for the internal lines. Joining the two external legs of the self-energy diagrams by a full Green function line gives us diagrams that appear in the perturbation series for $\Phi[G]$. There is also an overall normalization factor $1/\mathcal{N}$ associated with these diagrams (assuming spin-independent interactions) where $\mathcal{N}$ is the number of vertices, but details of this normalization factor are not central to our arguments here. The only requirement is the existence of $\Phi[G]$ whose derivative with respect to G gives us the proper self-energy (with all the correct normalization factors). Writing $\delta\Phi[G] = \int_p \Sigma(p)\delta G(p)$, and making use of Eq. (9.47), we find that the second contribution to the integral in Eq. (9.46) is

$$\int (dp)\Sigma(p)\frac{\partial}{\partial\epsilon}G(p) = \int (dp)\frac{\delta\Phi}{\delta G(p)}\frac{\partial}{\partial\epsilon}G(p). \tag{9.48}$$

The RHS in Eq. (9.48) is the change in $\Phi[G]$ when the frequencies on all its Green function lines are shifted by a fixed amount. Since at each vertex the total incoming frequency is exactly balanced by the total outgoing frequency, such a shift has no effect on $\Phi[G]$, and therefore the integral vanishes. As a result, Eq. (9.46) simplifies to

$$\begin{aligned} n &= i\int \frac{d\epsilon\, d^d\mathbf{p}}{(2\pi)^{d+1}}\, e^{i\epsilon\eta}\frac{\partial}{\partial\epsilon}\ln G(p) \\ &\equiv -i\int \frac{d\epsilon\, d^d\mathbf{p}}{(2\pi)^{d+1}}\, e^{i\epsilon\eta}\frac{\partial}{\partial\epsilon}\left[G(p)\frac{\partial G^{-1}(p)}{\partial\epsilon}\right]. \end{aligned} \tag{9.49}$$

We perform first the ϵ integration. Note that $G(p) = G^R(p)$ for $\epsilon > 0$ and $G^A(p)$ for $\epsilon < 0$ where $G^{R(A)}$ refers to the retarded (advanced) Green function. $G^R(p)$ is analytic everywhere in the upper half of the complex ϵ-plane, while G^A is analytic in the lower half. The frequency integration then yields

$$\begin{aligned} \int_{-\infty}^{\infty}\frac{d\epsilon}{2\pi}e^{i\epsilon\eta}\frac{\partial}{\partial\epsilon}\ln G(p) &= \int_0^{\infty}\frac{d\epsilon}{2\pi}e^{i\epsilon\eta}\frac{\partial}{\partial\epsilon}\ln G^R(p) + \int_{-\infty}^{0}\frac{d\epsilon}{2\pi}\frac{\partial}{\partial\epsilon}\ln G^A(p) \\ &= \int_{-\infty}^{\infty}\frac{d\epsilon}{2\pi}e^{i\epsilon\eta}\frac{\partial}{\partial\epsilon}\ln G^R(p) + \int_{-\infty}^{0}\frac{d\epsilon}{2\pi}\frac{\partial}{\partial\epsilon}\ln\frac{G^A(p)}{G^R(p)}. \end{aligned} \tag{9.50}$$

On the RHS of Eq. (9.50), the first integral vanishes. This is because G^R is analytic everywhere in the upper half plane and we may deform the contour of integration to $\text{Im}\epsilon = \infty$. The integral vanishes because of the presence of the $e^{i\epsilon\eta}$ factor. Since the retarded and advanced Green functions are complex conjugates of each other, $\ln(G^A/G^R)$ appearing in the integral is simply $i2\phi$, with ϕ being the phase of G^A. Thus

$$\int_{-\infty}^{\infty} \frac{d\epsilon}{2\pi} \frac{\partial}{\partial \epsilon} \ln G(p) = \frac{i}{\pi}[\phi(0, \mathbf{p}) - \phi(-\infty, \mathbf{p})]. \tag{9.51}$$

In going from $\epsilon = 0$ to $\epsilon = -\infty$, the phase difference $\phi(0) - \phi(-\infty)$ is finite (and equal to $-\pi$) provided the quasiparticle energy $\xi(\mathbf{p}) \approx \xi^{(0)}(\mathbf{p}) + \text{Re } \Sigma(\xi(\mathbf{p}), \mathbf{p})$ is negative (for that ensures a pole of G^A in the $\epsilon < 0$ half-plane and crossing this pole along the negative real axis advances the phase by π). For positive values of the quasiparticle energy, $\xi^{(0)}(\mathbf{p}) + \text{Re } \Sigma(\xi(\mathbf{p}), \mathbf{p}) > 0$, the phase difference is zero because the pole of G^A lies in the $\epsilon > 0$ half-plane. We thus arrive at Luttinger's theorem,

$$n = \int (d\mathbf{p}) \Theta(G^A(0, \mathbf{p})), \tag{9.52}$$

where we made use of the fact that for negative quasiparticle energies, $G^A(0, \mathbf{p}) > 0$. Note that $\xi(\mathbf{p})$ vanishes at $|\mathbf{p}| = p_F$, which means that the momentum integral in Eq. (9.52) extends up to the Fermi momentum p_F. Since p_F remains unchanged by momentum conserving interactions, and so does the number density n, we conclude that the volume enclosed by the Fermi surface in momentum space is unchanged by electron interactions. For the spinful case, an additional factor of 2 appears on the RHS of Eq. (9.52) on account of the spin degeneracy.

What happens to Luttinger's theorem if the Green function also has zeros apart from poles? In that case, instead of the integral in Eq. (9.49) just counting the poles of $G^A(\epsilon, \mathbf{p})$ on the half-line $-\infty < \epsilon < 0$, (see Eq. (9.50), we can see using the argument theorem of contour integrals, that the RHS of Eq. (9.49) represents the difference of the number of poles and number of zeros of the Green function.

9.5 Stoner instability and transition to ferromagnetic metal phase

In Sec. 2.2, we showed that in the presence of a short-range repulsive interaction of electrons of strength I, the electron gas is unstable against a transition to a ferromagnetic metal phase when $I\nu(0) > 1$. The short range nature of the interaction makes it possible for us to discuss such metallic ferromagnetism within the context of Landau theory. In the Landau phenomenology, the quantity $-I\nu(0)$ is identifiable as the parameter F_0^a. Before getting into the Landau formalism, we first make a simpler analysis along the lines of Chapter 6 where we studied the response of an ideal Fermi gas.

In the literature, the mechanism of metallic ferromagnetism discussed below goes by the name "Stoner ferromagnetism". Here, the ferromagnetism is the result of balancing excess kinetic energy of spin-polarized conduction electrons with interaction (potential) energy gain in the presence of the polarization. This is different from ferromagnetism of localized spins (e.g. the Heisenberg model), in which the kinetic energy of the spins is basically quenched. Stoner ferromagnetism is thus very relevant for metallic ferromagnets such as iron and nickel. As the repulsive exchange interaction I leading to Stoner ferromagnetism is of the form of a local correlation, it is more likely to be realized in partially filled d and f electron bands that are relatively narrow (i.e. smaller kinetic energies than s or p bands) and the electrons tend to be more localized.

The Stoner instability is a continuous phase transition since, as we will see below, the magnetization m develops continuously from zero at $I\nu(0) = 1$. We consider small magnetizations, $m/n \ll 1$ and low temperatures $T \ll \mu(0)$, where $\mu(0) = \epsilon_F + In$ is the chemical potential at zero temperature and magnetization. For finite m, T, we denote the chemical potential by $\mu(m,T) \approx \mu(0) + \delta\mu(m,T)$. The self-consistent equations for m and n obtained in Sec. 2.2 are rewritten as

$$n = \frac{1}{2}\int d\epsilon[\rho(\epsilon + Im) + \rho(\epsilon - Im)]n_\epsilon, \tag{9.53}$$

$$m = \frac{1}{2}\int d\epsilon[\rho(\epsilon + Im) - \rho(\epsilon - Im)]n_\epsilon. \tag{9.54}$$

$\rho(\epsilon)$ is the electron density of states at energy ϵ (including the spin degeneracy factor of 2) and $n_\epsilon = 1/[\exp((\epsilon - \mu)/T) + 1]$ is the Fermi-Dirac distribution function. We expand the density of states and the distribution function in both the equations in a power series in m, and $\delta\mu$:

$$n \approx \int_0^\infty d\epsilon \left[\rho(\epsilon) + \rho''(\epsilon)(Im)^2/2\right] n_\epsilon, \tag{9.55}$$

$$m \approx \int_0^\infty d\epsilon \left[\rho'(\epsilon)(Im) + \rho'''(\epsilon)(Im)^3/6\right] n_\epsilon. \tag{9.56}$$

For the low temperature behavior, we use Sommerfeld's expansion for integrals involving Fermi-Dirac distribution functions,

$$F(\eta) = \int_0^\infty \frac{dx\,\phi(x)}{\exp(x-\eta)+1} \approx \int_0^\eta dx\,\phi(x) + \frac{\pi^2}{6}\phi'(\eta) + \frac{7\pi^4}{360}\phi'''(\eta) + \ldots. \tag{9.57}$$

We thus obtain

$$n \approx n + \nu(0)\delta\mu + \left[\frac{\pi^2}{6}T^2 + \frac{1}{2}(Im)^2\right]\nu'(0), \tag{9.58}$$

$$m \approx (Im)\left[\nu(0) + \frac{\pi^2}{6}T^2\nu''(0) + \frac{1}{6}(Im)^2\nu''(0) + \delta\mu\nu'(0)\right]. \tag{9.59}$$

We used the fact that $\int_0^{\mu(0)} d\epsilon\rho(\epsilon) = n$, and the notation $\rho(\mu(0)) \equiv \nu(0)$. Using Eq. (9.58) to eliminate $\delta\mu$ from Eq. (9.59), we arrive at the following equations for determining m :

$$m = A(T)m + Bm^3, \tag{9.60}$$

$$A(T) = I\nu(0)\left[1 - \frac{\pi^2}{6}T^2\left\{\left(\frac{\nu'(0)}{\nu(0)}\right)^2 - \left(\frac{\nu''(0)}{\nu(0)}\right)\right\}\right], \tag{9.61}$$

$$B = -I^3\nu(0)\left[\frac{1}{2}\left(\frac{\nu'(0)}{\nu(0)}\right)^2 - \frac{1}{6}\left(\frac{\nu''(0)}{\nu(0)}\right)\right]. \tag{9.62}$$

At low temperatures, clearly $A > 0$, in which case, a nontrivial solution for m may exist only if $B < 0$. It is given by $m = \sqrt{(1-A)/B}$, which is real only if $A \geq 1$. For $I\nu(0 > 1$, the ferromagnetic phase may exist at finite temperatures provided the temperature is sufficiently low, and the temperature T_C for the transition (Curie temperature) is given by the solution of $A(T_C) = 1$. It is easy to see that

$$T_C = C\sqrt{1 - 1/I\nu(0)}, \tag{9.63}$$

where

$$C^{-2} = \frac{\pi^2}{6}\left[\left(\frac{\nu'(0)}{\nu(0)}\right)^2 - \left(\frac{\nu''(0)}{\nu(0)}\right)\right]. \tag{9.64}$$

Since $C \sim \mu(0)$, Stoner theory predicts that the transition temperature is of the order of the Fermi energy – a fact belied by actual transition temperatures of ferromagnetic metals such as iron ($T_C = 1043$K and Fermi energy $\sim 10^5$K). The Stoner instability at zero temperature is an example of a quantum phase transition.

Experimentally, $I\nu(0)$ can be tuned by the application of hydrostatic pressure or chemical doping, through their effect on the density of states as well as electron interaction strength. Indeed, in ferromagnetic metals such as the intermetallic UGe_2[11] and the *d*-band metal $ZrZn_2$[12] with small Curie temperatures, application of hydrostatic pressure results in a ferromagnet to paramagnet transition.

[11]S. S. Saxena, P. Agarwal, K. Ahilan, F. M. Grosche, R. K. W. Haselwimmer, M. J. Steiner, E. Pugh, I. R. Walker, S. R. Julian, P. Monthoux, G. G. Lonzarich, A. Huxley, I. Sheikin, D. Braithwaite and J. Flouquet, *Nature* **406**, 587 (2000).

[12]C. Pfleiderer, M. Uhlarz, S. M. Hayden, R. Vollmer, H. v. Löhneysen, N. R. Bernhoeft and G. G. Lonzarich, *Nature* **412**, 58 (2001).

In the paramagnetic phase, a finite magnetization can be induced by applying an external magnetic field $B_{\rm ext}$. This has the effect of shifting the quasiparticle energies, $\epsilon_{\uparrow(\downarrow)} \to \epsilon \pm \mu_B B_{\rm ext}$, or in Eq. (9.54), a shift $Im \to Im + \mu_B B_{\rm ext}$. For small fields, we may ignore any change in the chemical potential. From Eqs. (9.60) and (9.61) it then follows that $m \approx (A/I)(Im + \mu_B B_{\rm ext})$. Thus,

$$m(T) \approx \frac{A(T)}{1 - A(T)} \mu_B B_{\rm ext} \equiv \frac{(A(T)/I)}{A(T_{\rm C}) - A(T)} \mu_B B_{\rm ext}, \tag{9.65}$$

where $A(T) < 1$ in the paramagnetic phase and we used the relation $A(T_{\rm C}) = 1$. Expressing the induced magnetic moment in terms of the uniform magnetic susceptibility, $\chi(T)B_{\rm ext} \equiv \mu_B m(T)$, we have $\chi(T) = (A/I)\mu_B^2/(1-A)$. We use the expression for $A(T)$ obtained in Eq. (9.65) to arrive at our final expression for the temperature dependence of the uniform magnetic susceptibility. At zero temperature, we get $\chi(T) = \mu_B^2 \nu(0)/[1 - I\nu(0)]$, in agreement with Landau theory. In the vicinity of the Curie temperature:

$$\chi(T) \approx \mu_B^2 \frac{(C^2/I^2\nu(0))}{T^2 - T_{\rm C}^2}, \tag{9.66}$$

which agrees with the well-known Curie-Weiss law provided $|T/T_{\rm C} - 1| \lesssim 1$, but deviates appreciably at higher temperatures. The temperature dependence of magnetization for $T \lesssim T_{\rm C}$ can be similarly obtained:

$$\begin{aligned} m(T) &= \sqrt{(A(T) - 1)/|B|} \equiv \sqrt{[A(T) - A(T_{\rm C})]/|B|} \\ &= \sqrt{\frac{I\nu(0)}{|B|C^2}} \sqrt{T_{\rm C}^2 - T^2} \simeq \sqrt{T_{\rm C} - T}, \end{aligned} \tag{9.67}$$

with the square root growth of the order parameter a signature of our mean-field treatment (ignoring thermal and quantum fluctuations of magnetization).

While the Stoner mechanism does provide a physical mechanism of ferromagnetism in metals, and even offers an explanation for the Curie-Weiss behavior of the magnetic susceptibility in the vicinity of the Curie temperature, it still suffers from two key shortcomings. First, as we have already noted, the predicted Curie temperature in Stoner theory turns out to be comparable to the Fermi energy, which is much larger than actual values of the Curie temperature in metallic ferromagnets. Second, as is evident from Eq. (9.66), the Curie-Weiss law holds in the immediate neighborhood of $T_{\rm C}$ but shows appreciable deviations at higher temperatures. Such issues are a consequence of neglecting the effect of low-energy magnetization fluctuations on the properties near the critical temperature.

9.5.1 Relation to microscopics of Landau Fermi liquid theory

In this section, we will study the effect of electron density fluctuations on the Stoner instability, particularly to understand their effect on the mean-field condition, $I\nu(0) = 1$, for the instability. We will use the diagrammatic perturbation theory for the Landau Fermi liquid (introduced earlier in this chapter) for this purpose. Since we are discussing a ferromagnetic instability, we now need to consider the full spin-dependent vertex functions, $\Gamma_{\alpha\beta;\gamma\delta}(p, p'; p+q, p'-q)$, for small q. As before, the spin indices are taken to be in the same order as the momenta (p and p' are the incoming momenta and α, β their respective spins). We have already seen for the spinless case that the vertex function in the ω-limit is related to the Landau f-function through $f = \mathcal{Z}^2\Gamma^\omega$, or $F = \nu(0)\mathcal{Z}^2\Gamma^\omega$. The same relation holds for the spinful case. For isotropic systems with time reversal symmetry, the spin dependence of the vertex function appears as the scalar product of the Pauli matrices, $\boldsymbol{\sigma}$ and $\boldsymbol{\sigma}'$, and the momentum dependence is only through the angular separation of the momenta of the two scattering particles (see discussion following Eq. (1.28) for an explanation). The instabilities and collective excitations are determined by the poles of the corresponding response (or, equivalently, vertex) functions – for example, low energy magnetic excitations are determined by the poles of the magnetic susceptibility. For spin conserving interactions, the χ^ω part of the spin susceptibility vanishes, and we just need to consider the singular (quasiparticle) part of the response function.

Consider first the uniform and static susceptibility, which corresponds to the $\mathbf{q}$-limit of the response (or vertex) functions. Let us write

$$\nu(0)\mathcal{Z}^2\Gamma^\omega_{\alpha\beta,\gamma\delta}(\mathbf{p}, \mathbf{p}') = F^s(\mathbf{p}, \mathbf{p}')\delta_{\alpha\gamma}\delta_{\beta\delta} + (\boldsymbol{\sigma}_{\alpha\gamma} \cdot \boldsymbol{\sigma}'_{\beta\delta})F^a(\mathbf{p}, \mathbf{p}'), \tag{9.68}$$

$$\mathcal{Z}^2\Gamma^{\mathbf{q}}_{\alpha\beta,\gamma\delta}(\mathbf{p}, \mathbf{p}') = T^s(\mathbf{p}, \mathbf{p}')\delta_{\alpha\gamma}\delta_{\beta\delta} + (\boldsymbol{\sigma}_{\alpha\gamma} \cdot \boldsymbol{\sigma}'_{\beta\delta})T^a(\mathbf{p}, \mathbf{p}'). \tag{9.69}$$

In the second equation, we made use of the relation, Eq. (9.30), between the vertex function in the $\mathbf{q}$-limit and the T-matrix (or scattering amplitude) for excitations near the Fermi surface. Note also that $\nu(0) = m^* p_F/\pi^2$ is the density of states including the Fermi liquid corrections. We use the above relations in the Bethe-Salpeter equation, Eq. (9.25), (generalizing it to the spinful case). The spin-symmetric and spin-antisymmetric parts separate and we get

$$\nu(0)T^s(\mathbf{p}, \mathbf{p}') = F^s(\mathbf{p}, \mathbf{p}') - \nu(0)\int \frac{d\Omega_{\mathbf{k}}}{4\pi} F^s(\mathbf{p}, \mathbf{k})T^s(\mathbf{k}, \mathbf{p}'), \tag{9.70}$$

$$\nu(0)T^a(\mathbf{p}, \mathbf{p}') = F^a(\mathbf{p}, \mathbf{p}') - \nu(0)\int \frac{d\Omega_{\mathbf{k}}}{4\pi} F^a(\mathbf{p}, \mathbf{k})T^s(\mathbf{k}, \mathbf{p}'). \tag{9.71}$$

The T and F functions depend only on the angle between the vectors. We denote by θ the angle between $\mathbf{p}$ and $\mathbf{p}'$, and expand the T and F functions in a Legendre series such as $F^s(\theta) = \sum_{l=0}^{\infty} F_l^s P_l(\cos\theta)$. Substituting in Eq. (9.71) we readily obtain the solutions for the T-functions in terms of the F-functions:

$$T_l^{s(a)} = \frac{1}{\nu(0)} \frac{F_l^{s(a)}}{1 + \frac{F_l^{s(a)}}{2l+1}}. \tag{9.72}$$

We immediately see that the condition for instability in Eq. (9.72), is the same as the Pomeranchuk conditions $F_l^{s(a)} = -(2l+1)$. The Stoner ferromagnetic transition is a special instance of a Pomeranchuk instability corresponding to $F_0^a = -1$.

The remaining task is to express the F functions, or equivalently the Γ^ω vertex, in some sort of a perturbation expansion. In the Stoner model, we have a local repulsive interaction between electrons of opposite spins, $H_{\text{int}} = I\sum_\sigma \int(d\mathbf{p})(d\mathbf{p}')n_{\mathbf{p}}(\sigma)n_{\mathbf{p}'}(\bar{\sigma})$. The leading term in the expansion of $\Gamma^\omega(p,p')$ (and also the irreducible vertex $I(p,p')$ is

$$\begin{aligned}\Gamma^{\omega,1}(p,p')_{\alpha\beta,\gamma\delta} \equiv I^{(1)}_{\alpha\beta,\gamma\delta}(p,p') &= I(\delta_{\alpha\gamma}\delta_{\beta\delta} - \delta_{\alpha\delta}\delta_{\beta\gamma}) \\ &\equiv \frac{I}{2}\delta_{\alpha\gamma}\delta_{\beta\delta} - \frac{I}{2}\boldsymbol{\sigma}_{\alpha\gamma}\cdot\boldsymbol{\sigma}'_{\beta\delta}. \end{aligned} \tag{9.73}$$

Here the superscript 1 refers to the order in the perturbation expansion, σ and σ' are Pauli matrices associated with the spins of the scattering particles, and we obtained the spin symmetric (antisymmetric) component of the interaction by taking the sum (difference) of interaction energies with parallel and antiparallel spin configurations. An alternate way is to use the so-called completeness relation of the Pauli spin matrices,

$$\boldsymbol{\sigma}_{\alpha\gamma}\cdot\boldsymbol{\sigma}'_{\beta\delta} = 2\delta_{\alpha\delta}\delta_{\beta\gamma} - \delta_{\alpha\gamma}\delta_{\beta\delta}. \tag{9.74}$$

Clearly, $\Gamma^{\omega,1}_{\alpha\beta,\gamma\delta}$ is antisymmetric with respect to interchange of spin labels on the incoming or outgoing legs. Diagrams corresponding to these two contributions to $\Gamma^{\omega,1}$ are shown in Fig. 9.4(a) and (b). The spin-dependent part depends only on the dot product of the spin operators of the scattering particles – a reflection of isotropy and time-reversal symmetries in our model. The spin-symmetric and spin-antisymmetric components have the same magnitude and opposite signs – this is a consequence of the local nature of the interaction we have chosen. In comparison, we had observed for long-range Coulomb interactions in Sec. 6.3.2, that case both components were equal. The reason is not difficult to see. From Fig. (9.4)(a) and (b) it is evident that for a general two-particle interaction U of the central-force kind, $I^{(1)}_{\alpha\beta,\gamma\delta}(p,p';p,p') = U(0)\delta_{\alpha\gamma}\delta_{\beta\delta} - U(\mathbf{p}-\mathbf{p}')\delta_{\alpha\delta}\delta_{\beta\gamma}$. Now $U(\mathbf{q}\to 0)$, which is governed by long-distance behavior of the Coulomb interaction, is divergent, but in overall charge neutral systems, this term is exactly cancelled by a negative contribution from long-range electron interaction with the background positive charges. Removal of $U(0)$ makes the spin-symmetric and spin-antisymmetric components of $I^{(1)}$ equal.

The mean field treatment of the Stoner instability corresponds to approximating Γ^ω by its leading term in the perturbation expansion, $\Gamma^{\omega,1}$. Thus using $F_0^{s(a)} \approx \pm\nu(0)I$ in Eq. (9.72) (disregarding the Fermi-liquid quasiparticle weight), we obtain the following solution for the T-matrices:

$$T_0^{s(a)} = \pm\frac{I\nu(0)/2}{1 \pm I\nu(0)/2}, \tag{9.75}$$

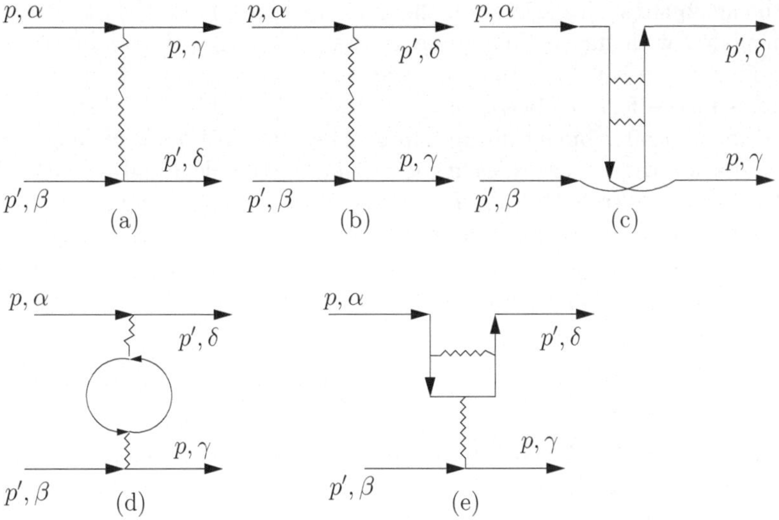

FIGURE 9.4: Diagrams comprising the irreducible vertex function $I(p, p')$ contributing to the vertex function $\Gamma^{\omega}_{\alpha\beta,\gamma\delta}(p, p')$ up to "second" order in the perturbation expansion. The wavy lines represent the bare interaction I and the solid lines represent the full Green functions. The internal Green function lines contain a sum over all possible irreducible self-energy insertions. Diagrams (c)-(e) contain two-particle propagators transferring a net momentum of $p - p'$ from the top G line to the bottom one, and may involve ladder type [(c)], bubble type [(d)], or contain elements of both ladders and bubbles [(e)]. In accordance with diagram rules, each fermionic bubble is associated with an extra negative sign; thus (d) and (e) cancel out. Diagrams with two particle Green functions of the particle-particle type carrying a total momentum $p+p'$ are not being shown (see text).

which gives us the mean-field condition for Stoner ferromagnetism. The instability condition is quantitatively different from the earlier $I\nu(0) = 1$; this is an artifact of the different mean-field approximations in the two cases. In the present scheme, considered full rotational symmetry of the spins while in the earlier mean-field analysis, a particular direction (i.e. z) was chosen for the magnetization.

Let us now examine the "second" order contributions[13]. In Fig. 9.4, diagrams (c), (d) and (e) are some that appear in the next to leading order of the diagrammatic expansion of the irreducible vertex, $I(p, p')$. In these terms, a net four-momentum $p - p'$ is transferred from the upper line to the lower one, and they are of the same type as Fig. 9.1(b), with the bare vertex at

[13]The quotation marks are there because the internal lines in these diagrams are full Green functions which include irreducible self-energy insertions to all orders in I.

both the squares. In Fig. 9.4(c), the transfer of $p - p'$ involves a two-particle propagator with interactions appearing as two rungs of a "ladder". In (d), the momentum is transferred via a sequence of strings and two-particle propagators in the form of beads or "bubbles". The bubble involves summation over an internal momentum and spin. The standard rules of diagrammatic perturbation theory accord each such fermionic bubble a multiplicative factor of (-1) to the overall diagram. Diagrams of the type (e) involve a combination of a string and a ladder (there are two such diagrams depending on whether the string is attached to the upper or lower line). Clearly, (d) and (e) cancel each other out leaving only (c) as the sole contributor at second order in Fig. 9.4. Thus,

$$I^{(2)}_{\alpha\beta,\gamma\delta}(p,p') = -\frac{I^2}{2}\Pi(p-p')\delta_{\alpha\gamma}\delta_{\beta\delta}, \tag{9.76}$$

and $\Pi(p-p') = 2\int(dq')G(q')G(p'-p+q')$ is essentially the Lindhard polarizability Π (written without the conventional multiplicative factor e^2) inclusive of the spin degeneracy factor of 2 in line with the convention we have been following. In the limit of zero momentum transfer, $(p_0 - p'_0) = 0$, $|\mathbf{p}-\mathbf{p}'| \to 0$, the polarizability $\Pi = (-1)\mathcal{Z}^2 \times \nu(0)$.

The second order contribution to the irreducible vertex, $I^{(2)}_{\alpha\beta,\gamma\delta}$, that we have calculated, does not have the fully antisymmetric spin structure of $I^{(1)}_{\alpha\beta,\gamma\delta}$ under interchange of $\alpha \leftrightarrow \beta$ or $\gamma \leftrightarrow \delta$. We will show later that there are other contributions in the second order where the spin structure is of the form $\delta_{\alpha\delta}\delta_{\beta\gamma}$. Whether these and other omitted diagrams restore the fully antisymmetric spin structure is a more subtle question that we do not intend to resolve in this book, but below we will mention appropriate literature where this issue has been addressed.

We note that diagrams (a) and (b) can be interpreted as the leading terms of the ladder-type and bubble-type series. The remaining particle-hole type terms correspond in general to mixed ladder and bubble diagrams. The key differences in the structure of a ladder and a bubble are (i) the two propagators in a bubble carry the same spin while in a ladder all spin combinations are possible, and (ii) each bubble comes with an extra multiplicative factor of (-1).

Two kinds of terms that appear in Γ^ω in the second order are missing from Fig. 9.4. First, there are diagrams where the internal lines convey a four-momentum $p+p'$ from the upper to the lower line, and are of the same type as Fig. 9.1(a). The internal Green functions in this case are of the particle-particle type and become singular in the limit of vanishing total four-momentum, $k = p + p'$, of the scattering particles. Up to second order in the perturbation theory, these processes do not couple to the particle-hole channel. Near a magnetic instability, the particle-hole vertex in the spin channel is large. Therefore in the literature, one often finds that particle-hole diagrams of ladder and bubble type are summed up to infinite order, neglecting the coupling with particle-particle type propagators at third and higher order. This is what we will also do. But a formally correct treatment should include the corrections from particle-particle channels, especially if it is established that inserting particle-particle vertices and summing over the independent internal coordinates associated with such vertices gives contributions comparable to the particle-hole vertices at the *same* order in the perturbation expansion.

The second type of diagrams not shown in Fig. 9.4 are those having the structure of Fig. 9.1(c) but containing only the nonsingular part, R of the internal particle-hole propagator (see Eq. (9.16)). The nonsingular part contains all contributions to the small momentum particle-hole propagator $G(q+q')G(q')$ save those coming from the immediate vicinity of the the closely spaced (quasiparticle) poles of the two Green functions. The nonsingular part R vanishes in the limit of quasiparticle weight $\mathcal{Z} \to 1$. The reason why we have such diagrams comes from the very definition of Γ^ω; for, $\Gamma^\omega(p,p') = I(p,p') - i\int(dq')I(p,q')R(q')\Gamma^\omega(q',p')$ not only includes the irreducible vertex $I(p,p')$ but also the contribution $-i\int(dq')I(p,q')R(q')\Gamma^\omega(q',p')$.

Let us now sum over irreducible vertices involving only particle-hole bubbles and ladders (referred hereafter as simply the ladder diagrams). The result is

$$I_{\alpha\beta,\gamma\delta}(p,p') \approx \frac{I}{1+I\Pi(p-p')/2}\delta_{\alpha\gamma}\delta_{\beta\delta} - \frac{I}{1-(I\Pi(p-p')/2)^2}\delta_{\alpha\delta}\delta_{\beta\gamma} \tag{9.77}$$

$$= -\frac{I/2}{1-I\Pi(p-p')/2}\delta_{\alpha\delta}\delta_{\beta\gamma} + \frac{I/2}{1+I\Pi(p-p')/2}\boldsymbol{\sigma}_{\alpha\delta}\cdot\boldsymbol{\sigma}'_{\beta\gamma}. \tag{9.78}$$

It is evident from Eq. (9.78) that local repulsive interactions enhance the irreducible vertex in the spin channel while suppressing it in the charge channel. In Eq. (9.78), the structure of the two terms, $\delta_{\alpha\delta}\delta_{\beta\gamma}$ and $\boldsymbol{\sigma}_{\alpha\delta}\cdot\boldsymbol{\sigma}'_{\beta\gamma}$ respectively, are different from the $\delta_{\alpha\gamma}\delta_{\beta\delta}$ and $\boldsymbol{\sigma}_{\alpha\gamma}\cdot\boldsymbol{\sigma}'_{\beta\delta}$ structure of the spin-symmetric and spin-antisymmetric components in Landau theory. At this stage one must exercise caution and not rush to read off a "Stoner criterion" for the ferromagnetic transition from the pole of the spin-dependent component of $I(p,p')$. As we have emphasized above, the Stoner criterion is obtained from the pole of the *full* vertex function Γ. Still, in the vicinity of the magnetic transition, we expect the second term in Eq. (9.78) to be significantly larger than the first, and we therefore approximate

$$I_{\alpha\beta,\gamma\delta}(p,p') \approx \frac{I/2}{1+I\Pi(p-p')/2}\boldsymbol{\sigma}_{\alpha\delta}\cdot\boldsymbol{\sigma}'_{\beta\gamma}$$

$$\equiv \frac{3}{2}\frac{I/2}{1+I\Pi(p-p')/2}\left[\delta_{\alpha\gamma}\delta_{\beta\delta} - \frac{1}{3}\boldsymbol{\sigma}_{\alpha\gamma}\cdot\boldsymbol{\sigma}'_{\beta\delta}\right]. \tag{9.79}$$

We have used the completeness relation of the Pauli spin matrices, Eq. (9.74) and rewritten $I(p,p')$ with the conventional structure of the spin-symmetric and spin-antisymmetric components of Landau theory. The expression obtained in Eq. (9.79) for the irreducible vertex is also our approximate result for Γ^ω. We are neglecting corrections from particle-particle ladders and also the non-singular bit of the particle-hole ladder of the type Fig. 9.1(c). With these approximations, for momenta very close to the Fermi surface (i.e., $p_0 = \xi(\mathbf{p}) = 0$, and $p_0' = 0$), we identify the parameters f^c and f^s in Landau theory in the vicinity of a ferromagnetic instability,

$$f^s(\mathbf{p}, \mathbf{p}') \approx \frac{3\mathcal{Z}^2}{2} \frac{I/2}{1 + I\Pi(\mathbf{p} - \mathbf{p}')/2}, \tag{9.80}$$

$$f^a(\mathbf{p}, \mathbf{p}') \approx -\frac{\mathcal{Z}^2}{2} \frac{I/2}{1 + I\Pi(\mathbf{p} - \mathbf{p}')/2} \boldsymbol{\sigma}_{\alpha\gamma} \cdot \boldsymbol{\sigma}'_{\beta\delta}. \tag{9.81}$$

Let us more closely examine the spin structure of different contributions in the perturbation expansion of $I(p, p')$ and $\Gamma^\omega(p, p')$. We have already observed that the second order and higher contributions to the irreducible vertex in the particle-hole channel are of the form of Fig. 9.1(b) which is a particle-hole type diagram with a net momentum transfer of $p - p'$ from the upper to the lower lines. The bare vertex is just its leading term, $2I(\delta_{\alpha\gamma}\delta_{\beta\delta} - \delta_{\alpha\delta}\delta_{\beta\gamma})$, and has the fully antisymmetric structure. The next order in the perturbation expansion of $I(p, p')$ gives us $(-i)I^{(1)}_{\alpha\eta',\eta\delta}\, GG\, I^{(1)}_{\eta\beta,\gamma\eta'} = (-4i)I(GGI)\delta_{\alpha\gamma}\delta_{\beta\delta}$, which does not have the antisymmetric structure. In the third order, it is easily seen that the contribution, once again, has the antisymmetric structure $(\delta_{\alpha\gamma}\delta_{\beta\delta} - \delta_{\alpha\delta}\delta_{\beta\gamma})$. Thus, the contributions that are an odd power of I have the symmetry $(\delta_{\alpha\gamma}\delta_{\beta\delta} - \delta_{\alpha\delta}\delta_{\beta\gamma})$, while the even power terms have the symmetry $\delta_{\alpha\gamma}\delta_{\beta\delta}$. This particle-hole "ladder" series when summed gives us Eq. (9.78) quite straightforwardly.

In contrast to terms generated by such particle-hole ladders, the particle-particle ladder generated by diagrams of the type Fig. 9.1(a) have a different spin structure. In the second order we have $iI^{(1)}_{\alpha\beta,\xi\xi'}\, GG\, I^{(1)}_{\xi\xi',\gamma\delta} = (8i)I(GGI)(\delta_{\alpha\gamma}\delta_{\beta\delta} - \delta_{\alpha\delta}\delta_{\beta\gamma})$. In fact, it is easy to see that every term in this particle-particle ladder series has the fully antisymmetric spin structure. Summing such a series makes sense when the particle-particle propagator is large. When the total four momentum of the scattering particles is small, the particle-particle propagator has logarithmic singularities in this small four-momentum.

Vertices with particle-particle propagators, when inserted appropriately in even order terms in the particle-hole ladder diagrams, result in diagrams with the full antisymmetric spin structure. An example is shown in Fig. 9.5(a), where insertion of a particle-particle diagram into one of the particle-hole type at second order makes the overall spin structure fully antisymmetric. Similarly inserting two particle-particle diagrams into an $I^{(1)}$ vertex as in Fig. 9.5(b) results in a vertex with the structure $\delta_{\alpha\delta}\delta_{\beta\gamma}$.

We cannot outright rely on insertion of particle-particle diagrams to provide us a $\delta_{\alpha\delta}\delta_{\beta\gamma}$ component to the irreducible vertex in the second order (see Eq. (9.76)) as these appear at higher order in the perturbation expansion. However, since the particle-particle ladder diagrams (see e.g. Fig. 9.1(a)) are singular in the limit of vanishing total four-momentum k of the scattering

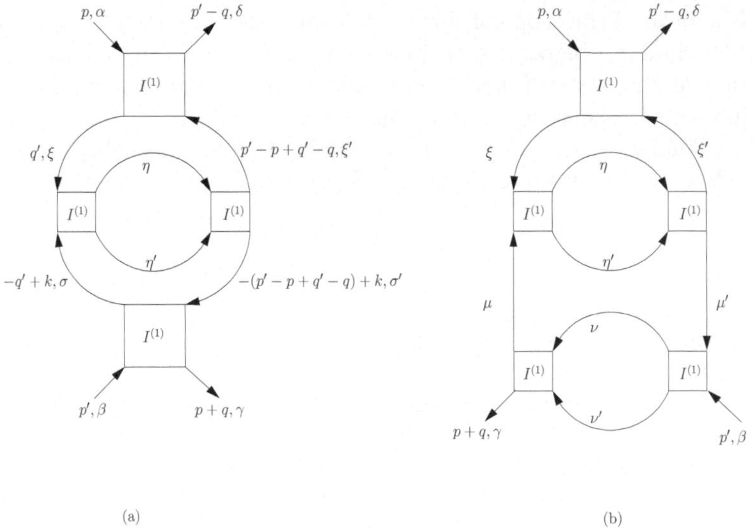

FIGURE 9.5: In (a) we show a fourth order diagram in the expansion of the irreducible vertex $I(p,p')$ showing a particle-particle correction to a particle-hole vertex. The bare vertices $I^{(1)}_{\alpha\beta,\gamma\delta}$ have a fully antisymmetric structure, $(\delta_{\alpha\gamma}\delta_{\beta\delta} - \delta_{\alpha\delta}\delta_{\beta\gamma})$. Upon tracing over the internal spin indices one finds that the overall spin structure of this vertex is also fully antisymmetric. In (b) we have a sixth order vertex with two particle-particle type vertices in an overall particle-hole diagram. The overall symmetry of this irreducible vertex is $\delta_{\alpha\delta}\delta_{\beta\gamma}$.

particles, they should be summed [14] if they do appear in this condition while integrating over the independent internal coordinates of the diagrams. Summing a series of particle-particle ladder diagrams at small $k = (k_0, \mathbf{k})$ gives us[15] the vertex $\Gamma_{pp}(p, -p+k, p+q, -p+k-q) \sim (2/\nu(0))/\ln(C/\max(k_0, v|\mathbf{k}))$, with C a (complex) constant with the dimensions of energy. While independent of the bare interaction, the diagram has a logarithmic singularity in the denominator whose value depends on the actual value of C (e.g. it is $\sim i\times$ Debye frequency if electron-electron interaction is mediated by phonons.). If C is of the order of the Fermi energy, then the large logarithmic factor in the denominator suppresses this vertex. On the other hand, if C has a small magnitude, then as $k \to 0$, one may encounter a pole in the vertex function.

[14] This was suggested in A. V. Chubukov and D. L. Maslov, *Phys. Rev. Lett.* **103**, 216401 (2009), where the authors proposed that inserting ladder summed particle-particle vertices into $I^{(2)}$ (as in Fig. 9.5(b)) generates the missing terms needed to restore the spin structure to that of the bare vertex at every order in the perturbation expansion of $I(p,p')$.

[15] See for example A. A. Abrikosov, L. P. Gorkov and I. E. Dzyaloshinskii, *Methods of Quantum Field Theory in Statistical Physics*, 2nd ed., Pergamon Press (1965).

It is beyond the scope of this book to provide a detailed account of the effects of including diagrams with particle-particle ladder insertions in the calculation of the vertex function near the Stoner instability, but we do refer the interested reader to the aforementioned literature.

Finally let us consider the remaining particle-hole diagrams that do not belong to the irreducible vertex $I(p,p')$ but nevertheless contribute to Γ^ω. These, as we have already mentioned, are the non-singular part of diagrams of the type Fig. 9.1(c) where the internal particle-hole propagator contains only the non-singular part R. The second order contribution of this kind is $-i\int(dq')I^{(1)}(p,q')R(q')I^{(1)}(q',p')$. Putting the spin labels on the irreducible vertices $I^{(1)}$, it is easily seen that this term has the overall spin structure $\delta_{\alpha\delta}\delta_{\beta\gamma}$, in contrast to $I^{(2)}(p,p')$ which is proportional to $\delta_{\alpha\gamma}\delta_{\beta\delta}$. Since we do not know the form of the function $R(k)$, we cannot directly compute such diagrams. So we write symbolically the Dyson equation $\Gamma^\omega = I - iIR\Gamma^\omega$, or in other words, $\Gamma^\omega = (1+iIR)^{-1}I$. One can view this as a further renormalization of the irreducible vertex due to the incoherent (i.e., non-quasiparticle) component of the Green functions.

9.6 Hertz theory

Recall that one of the key problems with Stoner mean field theory was that it predicted the ferromagnetic transition to take place at energy scales comparable to the Fermi energy. In Sec. 9.5 we studied some of the effects of quasiparticle interactions – in the spin channel – on the static properties of the Stoner ferromagnetic instability. In particular, we showed that quasiparticle interaction in the spin channel gets enhanced over the bare value due to corrections from low-energy particle-hole excitations. As a result, the ferromagnetic transition can now take place for values of I much less than the mean field prediction, $I\nu(0) = 1$. Such excitations in the spin channel, commonly referred to as *spin fluctuations* or *paramagnons* in the literature, affect not only the static properties of the Stoner instability but also dynamic ones. In this section we will study these effects in more detail.

We will take a different approach than before. So far we have focused on describing the ferromagnetic transition as an instability of the interacting electron system. An alternate approach, first developed by Hertz[16], attempts a description of the critical properties of Stoner ferromagnetism in terms of space and time-dependent order parameter field (here, the magnetization or spin fluctuation) whose dynamics is governed by an effective Landau-Ginzburg-Wilson (LGW) free energy functional, analytic in the order parameter field, and having the same global symmetries as the original fermion problem. Ferro-

[16] J. A. Hertz, *Phys. Rev.* B **14**, 1165 (1976).

magnetism is a broken symmetry state of this free energy functional. Such an approach is motivated by earlier developments in the theory of classical phase transitions in which, close to criticality, one first integrates out all short wavelength degrees of freedom except the long wavelength fluctuations of the order parameter. Taking this effective model as a starting point, one progressively eliminates short wavelength fluctuations obtaining successive effective models, resembling the earlier ones, but with renormalized parameters. In Hertz theory, quasiparticle degrees of freedom are formally integrated out, and their role is in determining the parameters in the starting free energy functional involving only the magnetization, or spin fluctuation degree of freedom. Although the classical free energy functionals do not involve time-dependent fields, it is tempting to imagine that the quantum problem is equivalent to the classical one at one dimension higher (this dimension being time). It turns out that the critical properties of the quantum model are not, in general, those of the corresponding classical one at one dimension higher, for, in critical quantum theories, physical quantities often behave differently with respect to time and space. Hertz showed that in d spatial dimensions, the quantum critical models have the same properties as corresponding classical critical ones in $d + z$ dimensions, where z, known as the dynamic exponent, is a number that depends on the scaling of low energy excitations with respect to the (small) wavevectors.

It is not immediately obvious why integrating out one set of low energy excitations such as the quasiparticles[17] while keeping the other (magnetization) will qualitatively simplify the problem. The fluctuating magnetization acts on the quasiparticles like a fluctuating Zeeman field. At the same time, the dynamics of the magnetic order parameter field is entirely generated by the quasiparticles themselves, since the order parameter, being a density, is bilinear in the quasiparticle fields and whose properties are those of the two-particle Green function. The quasiparticles generate an infinite number of n-body non-local interactions of the magnetization field. Unless it can be shown that the n-body interactions can be legitimately truncated at some reasonably small order, the entire scheme looks quite intractable to implement.

In general, these two types of excitations – quasiparticles and spin fluctuations – should be treated on the same footing. However considerable simplification is possible if the order parameter fluctuations (particle-hole type spin fluctuations) interacting with the quasiparticles do not fundamentally alter the validity of the perturbation theory underpinning the Landau Fermi liquid picture. To give an example of a possible breakdown of the Fermi liquid description, consider Fig. 9.1, where the diagrams of the type (a) and (b) are not singular in the limit of a vanishing transferred four momentum $q \to 0$, and were therefore incorporated into the irreducible vertex functions $I(p, p')$ and $\Gamma^{\omega}(p, p')$. Yet, it is evident that the particle-particle diagrams (type (a)) and

[17] It must be said that quasiparticles, while being low energy excitations, are also short wavelength, in contrast with the magnetization degree of freedom in Hertz theory that is both low energy and long wavelength.

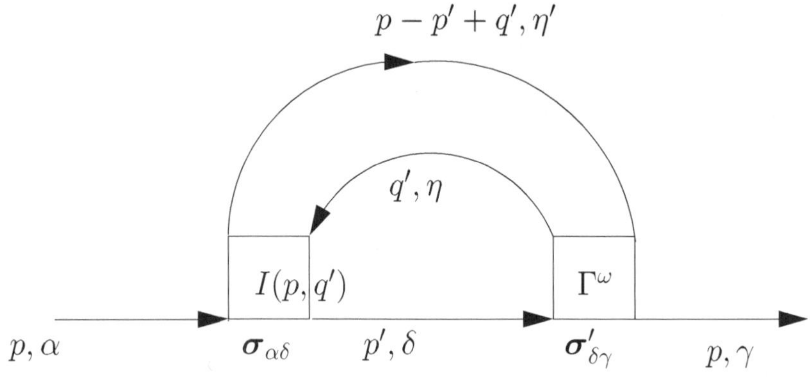

FIGURE 9.6: Diagram showing correction to the quasiparticle Green function by a spin fluctuation mode. The spin fluctuation (arch with the two squares at the ends and the Pauli matrices) carries a four momentum $p - p'$. The Pauli matrices at the vertices $I(p, q')$ and Γ^ω are there because of the projection to the spin channel. Since the spin fluctuation is a spin-1 boson, the incoming and outgoing spins can be equal or differ by ± 1.

particle-hole diagrams of the type (b) respectively become singular for small incoming four-momenta, $k = p + p' \to 0$, and for nearly identical incoming momenta, $p - p' \to 0$. If such vertices appear as part of self-energy diagrams with such a conspiracy of their incoming momenta, then the resulting singularities may ultimately show up as non-analytic terms in the LGW free energy functional of the magnetization field.

Note that in Sec. 9.5.1, a certain component of Γ^ω, defined now as a spin fluctuation, having the general form $V(p - p')\boldsymbol{\sigma}_{\alpha\delta} \cdot \boldsymbol{\sigma}'_{\beta\gamma}$, (see e.g. Eq. (9.81)) may be regarded as a spin-1 boson mediating an exchange interaction of quasiparticles with incoming four-momenta p and p' respectively. Figure 9.6 shows a correction to the quasiparticle Green function from coupling to a spin fluctuation excitation. In Hertz theory, pairs of quasiparticles are treated as the source of interaction of these spin fluctuations.

We now proceed to obtain the effective free energy functional of the order parameter field that governs the quantum critical behavior. As we are also ultimately interested in finite temperature properties[18] near the quantum critical point, we need to extend our diagram technique to finite temperatures. Thus we will employ the diagram technique for finite temperatures (known as the Matsubara formalism) which amounts to formulating quantum mechan-

[18] This will help address questions such as the effect of spin fluctuations on the temperature dependence of magnetic susceptibility, and how it differs from Stoner mean-field theory.

ics in imaginary time. The imaginary time τ extends over a finite interval, $0 \leq \tau \leq \beta$, where $\beta = 1/T$ is the inverse temperature.

9.6.1 Landau-Ginzburg-Wilson free energy functional

Let us begin as usual with the Stoner model introduced in Sec. 2.2. The first step in Hertz theory involves integrating out the short wavelength quasiparticle degrees of freedom. To this end we use a path-integral formulation which is completely equivalent to the diagrammatic perturbation theory we have been following. The advantage of the path integral formulation is that it provides an elegant way of integrating out the quasiparticle degrees of freedom, leaving behind an effective LGW free energy functional of the order parameter, or spin fluctuation, field. Observe that the model in Eq. (2.19) features an interaction term involving four fermions, so we cannot directly perform the integration over the fermion fields. However it is possible to make the model completely bilinear in the fermion fields, which can then be integrated out, by employing a Hubbard-Stratonovich transformation.

9.6.1.1 Hubbard-Stratonovich transformation

The basic idea behind this transformation is very simple. Consider an integral of the type

$$\mathcal{I} = \int_{-\infty}^{\infty} dx\, e^{-i\lambda^2 x^4 - i\frac{1}{2}x^2}, \tag{9.82}$$

involving a quartic polynomial in the exponent, and we assume λ is small. The quartic integral can be traded for a Gaussian one by introducing an auxiliary quantity, y :

$$\mathcal{I} = \mathcal{N} \int_{-\infty}^{\infty} dy \int_{-\infty}^{\infty} dx\, e^{-i\frac{1}{2}x^2 + i\lambda x^2 y - i\frac{1}{2}y^2}. \tag{9.83}$$

This is the Hubbard-Stratonovich transformation; in a nutshell, a completing the square exercise. Here $\mathcal{N}$ is a normalization factor. Think of x as the analogue of a fermionic field and y as a density field such as the magnetization, and our goal is to integrate out the x and obtain an effective distribution for y. It is now possible to formally perform the Gaussian integral over x, but note that we have ended up with another integral over y. Integration over x gives us (up to a normalization factor), $1/\sqrt{1-(2y\lambda)}$. Following this, we can make a perturbation expansion in λ and perform the y integral order by order. The result is an asymptotic series expansion of $\mathcal{I}$ in (even) powers of λ. We could also have directly expanded the integrand in Eq. (9.82) in a power series in a λ^2. Indeed, such a thing is what we do when we do perturbation expansions with four-fermion interactions using the path integral technique: the averaging is done with respect to the action of the unperturbed Hamiltonian which is analogous to the Gaussian weight in Eq. (9.82). Introducing the auxiliary quantity y offers its own advantages. Instead of averaging over x, (think fermions) we are often interested in averages involving y (think magnetization, and correlation functions of magnetization such as susceptibility). We rewrite the result following the x integration in Eq. (9.83) in the form

$$\begin{aligned} e^{-iF[y]} &= \mathcal{N} e^{-i\frac{1}{2}y^2 - \frac{1}{2}\ln[1-2y\lambda]} \\ &= \mathcal{N} e^{-i\frac{1}{2}y^2 + \frac{1}{2}\left[2y\lambda + \frac{1}{2}(2y\lambda)^2 + \frac{1}{3}(2y\lambda)^3 + \cdots\right]}. \end{aligned} \tag{9.84}$$

Since λ is small, we have retained only the first few terms in the logarithm in Eq. (9.84). The resulting exponential is an effective Gaussian theory (or equivalently a "free energy" $F[y]$) for y. Alternatively, we may also expand the free energy around its saddle point, $y^* : F'[y^*] = 0$, in which case the linear term no longer appears.

Let us now proceed to apply the Hubbard-Stratonovich transformation to our model, Eq. (2.19). We begin by rewriting our model again in the form of an action S in imaginary time (since we are going to use the path integral approach):

$$\begin{aligned} S[c] = S_0[c] + S_{\rm int}[c] &= V \int_0^\beta d\tau \int (d\mathbf{p}) \left(\sum_\sigma c^\dagger_{\mathbf{p},\sigma} \left[\partial_\tau + (\epsilon^{(0)}_{\mathbf{p}} - \mu) \right] c_{\mathbf{p},\sigma} \right) \\ &+ I \int_0^\beta d\tau \int d\mathbf{r} \sum_\sigma \left(c^\dagger_{\mathbf{r},\sigma} c_{\mathbf{r},\sigma} \right) \left(c^\dagger_{\mathbf{r},\bar\sigma} c_{\mathbf{r},\bar\sigma} \right). \end{aligned} \tag{9.85}$$

The second line of Eq. (9.85) is the interaction term, $S_{\rm int}$, $\beta = 1/T$ is the inverse temperature and V is the volume of the system. The normalization chosen is such that the fermion fields in the momentum representation are dimensionless. The interaction I has the dimensions of energy$\times$volume. In what follows, a degree of familiarity with the path integral approach will be helpful, since we do not intend to develop the technique here in great detail.

The partition function in the path integral formalism is

$$\mathcal{Z} = \mathcal{N} \int D[c]\, e^{-S[c]}, \tag{9.86}$$

with $\mathcal{N}$ a normalization factor whose details are not important to our analysis here, and the measure $D[c] \equiv Dc^\dagger Dc$ represents integration over all spatial and temporal configurations of the (fermionic) Grassman fields, subject to the Matsubara boundary condition for fermion fields, $c(\tau) = -c(\tau + \beta)$. The Grassman fields anticommute with each other, e.g., $c_{\mathbf{r},\sigma} c_{\mathbf{r}',\sigma'} = -c_{\mathbf{r}',\sigma'} c_{\mathbf{r},\sigma}$, $c^\dagger_{\mathbf{r},\sigma} c^\dagger_{\mathbf{r}',\sigma'} = -c^\dagger_{\mathbf{r}',\sigma'} c^\dagger_{\mathbf{r},\sigma}$ and $c^\dagger_{\mathbf{r},\sigma} c_{\mathbf{r}',\sigma'} = -c_{\mathbf{r}',\sigma'} c^\dagger_{\mathbf{r},\sigma}$.

As we discussed earlier in this chapter, we separate the interacting term in the charge and spin channel preserving the full rotational symmetry in spin space:

$$\sum_{\mathbf{r},\sigma} \left(c^\dagger_{\mathbf{r},\sigma} c_{\mathbf{r},\sigma} \right) \left(c^\dagger_{\mathbf{r},\bar\sigma} c_{\mathbf{r},\bar\sigma} \right) = \frac{1}{2} \sum_{\mathbf{r}} c^\dagger_{\mathbf{r},\alpha} c_{\mathbf{r},\gamma} c^\dagger_{\mathbf{r},\beta} c_{\mathbf{r},\delta} (\delta_{\alpha\gamma}\delta_{\beta\delta} - \boldsymbol{\sigma}_{\alpha\gamma} \cdot \boldsymbol{\sigma}_{\beta\delta}). \tag{9.87}$$

The RHS of Eq. (9.87) consists of two kinds of terms. First, there are terms of the type $c^\dagger_{\mathbf{r},\alpha}c_{\mathbf{r},\alpha}c^\dagger_{\mathbf{r},\beta}c_{\mathbf{r},\beta}$, which is just the square of the number density field at position $\mathbf{r}$. Second, there are terms of the type $(c^\dagger_{\mathbf{r},\alpha}\sigma^i_{\alpha\gamma}c_{\mathbf{r},\gamma})(c^\dagger_{\mathbf{r},\beta}\sigma^i_{\beta\delta}c_{\mathbf{r},\beta})$, which is the square of the magnetization density field whose components are $c^\dagger_{\mathbf{r},\alpha}\sigma^i_{\alpha\gamma}c_{\mathbf{r},\gamma}$, $i = 1, 2, 3$. We have seen earlier in our diagrammatic approach that near the ferromagnetic instability, repulsive short-range interactions of the quasiparticles suppress the charge susceptibility (and therefore charge fluctuations) while strongly enhancing the spin susceptibility. We therefore ignore spatial and temporal fluctuations of the number (charge) density field and replace it by a constant, the static expectation value.

The next step is the Hubbard-Stratonovich transformation. We are now left with interactions only in the spin channel. Observe that our action is a generalization of Eq. (9.82). We have an integral over the Grassman fields, $c^\dagger$, c, and the action contains both quadratic and quartic terms. The fields are continuous functions of τ. Despite all these complications, the procedure of introducing auxiliary fields to decouple the quartic interaction still works. For our purpose, the main difference between Gaussian integrals of the type $\mathcal{I} = \int \prod_i (dz_i^* dz_i) e^{-\sum_{ij} A_{ij} z_i^* z_j}$ over scalar fields (e.g. our example earlier in this section) and Grassman fields is that in the former, the result is $\mathcal{I} \propto 1/\text{Det } A$, while in the latter, $\mathcal{I} \propto \text{Det } A$ (Det A being the determinant of the matrix A). We introduce the auxiliary fields $\mathbf{m}(\mathbf{r},\tau)$ to decouple the quartic interaction term in the spin (magnetization density) channel[19]:

$$\mathcal{Z} = \mathcal{N} \int D[c, \mathbf{m}]\, e^{-S[c,\mathbf{m}]}, \tag{9.88}$$

where

$$\begin{aligned} S[c,\mathbf{m}] = V \int_0^\beta d\tau \Bigg\{ &\int (d\mathbf{p}) \left(\sum_\sigma c^\dagger_{\mathbf{p},\sigma} \left[\partial_\tau + (\epsilon^{(0)}_{\mathbf{p}} - \mu)\right] c_{\mathbf{p},\sigma} \right) \\ &- I \int d\mathbf{r} \left(\sum_{\alpha\gamma} \mathbf{m}(\mathbf{r},\tau) \cdot c^\dagger_{\mathbf{r},\alpha} \boldsymbol{\sigma}_{\alpha\gamma} c_{\mathbf{r},\gamma} - \frac{1}{2}\mathbf{m}(\mathbf{r},\tau)\cdot\mathbf{m}(\mathbf{r},\tau) \right) \Bigg\}. \end{aligned} \tag{9.89}$$

The magnetization fields satisfy the Matsubara boundary condition for bosons, $\mathbf{m}(\tau) = \mathbf{m}(\tau + \beta)$. We write the electron-magnetization coupling in

[19] More accurately, one should also decouple the quartic interaction in the charge channel, $(c^\dagger_{\mathbf{r},\alpha}c_{\mathbf{r},\alpha})(c^\dagger_{\mathbf{r},\beta}c_{\mathbf{r},\beta})$, by introducing another auxiliary field, $V(\mathbf{r},\tau)$, that has the meaning of the scalar potential. Thus $\exp[-(I/2)\int_{\tau,\mathbf{r}}(c^\dagger_{\mathbf{r},\alpha}c_{\mathbf{r},\alpha})(c^\dagger_{\mathbf{r},\beta}c_{\mathbf{r},\beta})] = \int D[V] \exp\left[-(1/2I)\int_{\tau,\mathbf{r}}(V(\mathbf{r},\tau))^2 - i\int_{\tau,\mathbf{r}} V(\mathbf{r},\tau)c^\dagger_{\mathbf{r},\alpha}c_{\mathbf{r},\alpha}\right]$. Observe the imaginary sign of the term coupling $V(\mathbf{r},\tau)$ to the charge density.

the momentum representation,

$$I \sum_{\alpha\gamma} \int d\mathbf{r}\, \mathbf{m}(\mathbf{r}) \cdot c^{\dagger}_{\mathbf{r},\alpha} \boldsymbol{\sigma}_{\alpha\gamma} c_{\mathbf{r},\gamma} = VI \sum_{\alpha\gamma} \int (d\mathbf{p})(d\mathbf{p}')(c^{\dagger}_{\mathbf{p},\alpha} c_{\mathbf{p}',\gamma})(\mathbf{m}_{\mathbf{p}-\mathbf{p}'} \cdot \boldsymbol{\sigma}_{\alpha\gamma}). \tag{9.90}$$

Following the Hubbard-Stratonovich decoupling, we now have a quadratic action for the fermionic fields. The fermion fields may now be integrated out but before that we express the time-dependent fields, c and $\mathbf{m}_{\mathbf{p}}$, in frequency space. The frequencies are discrete because of the finite time interval, and their values are such that the Matsubara boundary conditions on the fields, discussed above, are satisfied. We use the notation $k = (i\nu, \mathbf{k})$ to specify the (imaginary) frequency and momentum of a field:

$$c_{\mathbf{k}}(\tau) = \sum_{n=-\infty}^{\infty} c_k e^{-i\nu_n \tau}, \tag{9.91}$$

$$\mathbf{m}_{\mathbf{p}}(\tau) = \sum_{m=-\infty}^{\infty} \mathbf{m}_p e^{-i\omega_m \tau}. \tag{9.92}$$

For the fermionic fields c, the frequencies are $\nu_n = (2\pi/\beta)(n + 1/2)$, and for the bosonic fields $\mathbf{m}_{\mathbf{p}}$, the frequencies are $(2\pi/\beta)n$. In the four-momentum space (with $p = (i\nu_n, \mathbf{p})$, $p' = (i\nu_{n'}, \mathbf{p}')$, and $q = (i\omega_m, \mathbf{q})$), the action takes the form

$$S[c, \mathbf{m}] = V\beta \sum_{\alpha\gamma;n,n'} \int (d\mathbf{p})(d\mathbf{p}') \Big\{ (2\pi)^d \delta_{\alpha\gamma} \delta_{pp'} [-i\nu_n + (\epsilon^{(0)}_{\mathbf{p}} - \mu)] \\ -I\mathbf{m}_{p-p'} \cdot \boldsymbol{\sigma}_{\alpha\gamma} \Big\} c^{\dagger}_{p\alpha} c_{p'\gamma} + \frac{I\beta}{2} \sum_m \int (d\mathbf{q})\, \mathbf{m}_q \cdot \mathbf{m}_{-q}. \tag{9.93}$$

The δ-function in momentum and frequency space should be regarded as $\delta_{nn'}\delta(\mathbf{p} - \mathbf{p}')$. Integration over the fermionic fields in Eq. (9.88) gives us the determinant of the matrix in curly brackets in Eq. (9.93). We move the determinant to the exponent in the partition function. The effective action for the magnetization fields is then

$$S[\mathbf{m}] = -\mathrm{Tr} \Big[\ln \Big((2\pi)^d \delta_{\alpha\gamma} \delta_{pp'} [-i\nu_n + (\epsilon^{(0)}_{\mathbf{p}} - \mu)] - I\mathbf{m}_{p-p'} \cdot \boldsymbol{\sigma}_{\alpha\gamma} \Big) \Big] \\ + \frac{I\beta}{2} \sum_m \int (d\mathbf{q})\, \mathbf{m}_q \cdot \mathbf{m}_{-q}. \tag{9.94}$$

The trace in Eq. (9.94) is over frequency, momentum and spin labels. At this point, the effective action is formally exact, but not yet very useful for calculations. We write the matrix inside the logarithm in the form $-G_0^{-1} - I\mathbf{m}_{p-p'} \cdot \boldsymbol{\sigma}_{\alpha\gamma}$, where $G_0^{-1}(p\alpha, p'\gamma) = (2\pi)^d \delta_{\alpha\gamma} \delta_{pp'} [i\nu_n - (\epsilon^{(0)}_{\mathbf{p}} - \mu)]$ is the inverse of the non-interacting fermion Green function. In fact, the non-interacting fermionic Green function is diagonal in the four-momentum and spin spaces, so we will simply write it as $G_0(p) = 1/[i\nu_n - (\epsilon_{\mathbf{p}}^{(0)} - \mu)]$. We then note that $\ln(-G_0^{-1} - I\mathbf{m}_{p-p'} \cdot \boldsymbol{\sigma}_{\alpha\gamma})$ can be separated into magnetization dependent and magnetization independent parts, $\ln(-G_0^{-1}) + \ln(1 + IG_0\mathbf{m} \cdot \boldsymbol{\sigma})$. The first of these terms, being $\mathbf{m}$-independent, may be absorbed in a redefinition of the normalization factor in the partition function. The partition function takes the form

$$\begin{aligned} \mathcal{Z} &= \mathcal{N} \int D[\mathbf{m}]\, e^{-S[\mathbf{m}]} \\ &= \mathcal{N} \int D[\mathbf{m}] \exp\left(\mathrm{Tr}\left[\ln(1 + IG_0\mathbf{m} \cdot \boldsymbol{\sigma})\right] - \frac{I\beta}{2} \sum_m \int (d\mathbf{q})\, |\mathbf{m}_{i\omega_m, \mathbf{q}}|^2 \right). \end{aligned} \tag{9.95}$$

From the partition function of Eq. (9.95) we can easily obtain the mean-field condition for the magnetization in the Stoner model (see Eq. (9.54)) by varying the effective action with respect to the magnetization component at $q = (i\omega_m, \mathbf{q}) = 0$, and setting it to zero[20]. Varying with respect to the magnitude m of this component we get the mean-field condition

$$\mathrm{Tr}\left[\frac{1}{1 + IG_0\mathbf{m} \cdot \boldsymbol{\sigma}} G_0 \hat{\mathbf{m}} \cdot \boldsymbol{\sigma} \right] = \beta m. \tag{9.96}$$

The trace is taken over spin, (fermionic) Matsubara frequencies as well as the momentum of the electrons. Since the magnetization has only a single Fourier component, the matrix $1 + G_0\mathbf{m} \cdot \boldsymbol{\sigma}$ is easily inverted:

$$\frac{1}{1 + G_0\mathbf{m} \cdot \boldsymbol{\sigma}} = \frac{1}{1 + IG_0 m} \left(\frac{1 + \hat{\mathbf{m}} \cdot \boldsymbol{\sigma}}{2} \right) + \frac{1}{1 - IG_0 m} \left(\frac{1 - \hat{\mathbf{m}} \cdot \boldsymbol{\sigma}}{2} \right). \tag{9.97}$$

This, upon substitution in Eq. (9.96) and taking the trace over the spin degrees of freedom gives

$$m = \frac{1}{\beta} \mathrm{Tr} \left[\frac{1}{i\nu_n - (\epsilon_{\mathbf{p}}^{(0)} - \mu) + Im} - \frac{1}{i\nu_n - (\epsilon_{\mathbf{p}}^{(0)} - \mu) - Im} \right]. \tag{9.98}$$

We perform the trace over the Matsubara frequencies using the contour integral technique or otherwise. For fermionic frequencies, $\nu_n = (2\pi/\beta)(n + 1/2)$, it can be shown that $\frac{1}{\beta} \sum_n (i\nu_n - (\epsilon_{\mathbf{p}}^{(0)} - \mu))^{-1} = n(\xi_{\mathbf{p}}) = [\exp(\beta\xi_{\mathbf{p}}) + 1]^{-1}$,

[20] This represents a specific stationary condition of the action.

with $\xi_{\mathbf{p}} = \epsilon_{\mathbf{p}}^{(0)} - \mu$.[21] We use this result in Eq. (9.98) to obtain

$$m = \int (d\mathbf{p})[n(\xi_{\mathbf{p}} - Im) - n(\xi_{\mathbf{p}} + Im)],$$

which is just the Stoner mean field equation for magnetization. Once again, breaking the full rotational symmetry and selecting a particular direction for magnetic order has resulted in the Stoner condition $I\nu(0) = 1$ instead of $I\nu(0)/2 = 1$ we obtained in Eq. (9.75) keeping the full rotational symmetry of the spins.

9.6.1.2 Effective free energy

In this section we obtain an effective free energy in terms of the fluctuating magnetization field near the ferromagnetic transition. We approach the transition from the paramagnetic side so that the mean field magnetization is zero. We expand the effective action in Eq. (9.95) in increasing powers of $\mathbf{m}$ around the saddle point $\mathbf{m} = 0$:

$$S[\mathbf{m}] \approx S^{(2)}[\mathbf{m}] + S^{(4)}[\mathbf{m}], \tag{9.99}$$

where

$$\begin{aligned} S^{(2)}[\mathbf{m}] &= \frac{I\beta}{2} \int_q \left[1 + (I/\beta)\left(\int_p G_0(p+q)G_0(p)\right)\right] \mathbf{m}_q \cdot \mathbf{m}_{-q} \\ &= \frac{I\beta}{2} \int_q \left(1 + \frac{I}{2}\Pi(q)\right) \mathbf{m}_q \cdot \mathbf{m}_{-q}, \end{aligned} \tag{9.100}$$

and

$$\begin{aligned} S^{(4)}[\mathbf{m}] = \frac{I^4}{4} \sum_{ijkl} \mathrm{Tr}\left(\sigma^i\sigma^j\sigma^k\sigma^l\right) \int_{q_1,q_2,q_3} m^i_{q_1} m^j_{q_2} m^k_{q_3} m^l_{-q_1-q_2-q_3} \\ \times \int_p G_0(p)G_0(p+q_1)G_0(p+q_1+q_2)G_0(p+q_1+q_2+q_3). \end{aligned} \tag{9.101}$$

Here $\int_q = \sum_m \int (d\mathbf{q})$ etc. is shorthand for the sum over Matsubara frequencies associated with q and integration over the momentum *without* any factors of $1/\beta$. To perform the trace over the Pauli matrix product in Eq. (9.101), we consider the Pauli matrices pairwise and use $\sigma_k\sigma_l = \delta_{ij}\mathbf{I} + i\epsilon_{kln}\sigma_n$ for each pair. Only the terms proportional to the identity matrix contribute to the

---

[21] A more general result,

$$\frac{1}{\beta}\sum_n (i\nu_n - \xi_{\mathbf{p}})^{-k} = \frac{1}{(k-1)!}\frac{d^{(k-1)}n(\xi_{\mathbf{p}})}{d\xi_{\mathbf{p}}^{(k-1)}},$$

is easily obtained by repeated differentiation of the $k = 1$ case.

trace. The result is $\mathrm{Tr}(\sigma_i\sigma_j\sigma_k\sigma_l) = 2(\delta_{ij}\delta_{kl} - \epsilon_{ijn}\epsilon_{kln})$. The expression for $S^{(0)}[\mathbf{m}]$ simplifies to

$$S^{(4)}[\mathbf{m}] = \frac{I^4}{2}\int_{q_1,q_2\cdot q_3}[(\mathbf{m}_{q_1}\cdot\mathbf{m}_{q_2})(\mathbf{m}_{q_3}\cdot\mathbf{m}_{-q_1-q_2-q_3}) \\ -(\mathbf{m}_{q_1}\times\mathbf{m}_{q_2})\cdot(\mathbf{m}_{q_3}\times\mathbf{m}_{-q_1-q_2-q_3})] \\ \times\int_p G_0(p)G_0(p+q_1)G_0(p+q_1+q_2)G_0(p+q_1+q_2+q_3). \quad (9.102)$$

As we are interested in slow and long wavelength fluctuations of the magnetization (i.e. q_i, $i = 1,2,3$ are all small), the cross products themselves are small compared to the dot products. Following Hertz, we drop the cross product term from Eq. (9.102):

$$S^{(4)}[\mathbf{m}] \approx \frac{I^4}{2}\int_{q_1,q_2,q}(\mathbf{m}_{q_1}\cdot\mathbf{m}_{q_2})(\mathbf{m}_{-q_1-q}\cdot\mathbf{m}_{-q_2+q}) \\ \times\int_p G_0(p)G_0(p+q_1)G_0(p-q)G_0(p-q_2). \quad (9.103)$$

The remaining task is to perform the Matsubara sum and momentum integration over $p = (i\nu_n, \mathbf{p})$. This will result in a kernel that is dependent on the three remaining four-momenta, q_1, q_2, q, which generates non-local interactions of the four magnetization fields in both space and time. We do not intend to get into all the complications of this non-local interaction. Instead, we use the fact that the four-momenta q_1, q_2, q are all small, and hence, as an approximation, we set these to zero inside the Green functions in Eq. (9.103). This simplifies $S^{(4)}$ a lot, since, by going over to the real space and imaginary time space, one can immediately see that the interaction of the magnetization fields is local. We perform the Matsubara sum over ν_n first:

$$\frac{1}{\beta}\sum_n \frac{1}{(i\nu_n - \xi_{\mathbf{p}})^4} = \frac{1}{3!}\frac{d^3 n(\xi_{\mathbf{p}})}{d\xi_{\mathbf{p}}^3}, \quad (9.104)$$

and the easiest way to obtain this is by repeated differentiation of the relation, $(1/\beta)\sum_n (i\nu_n - \xi_{\mathbf{p}})^{-1} = n(\xi_{\mathbf{p}})$. Consequently,

$$\sum_n \int (d\mathbf{p}) G_0^4(i\nu_n, \mathbf{p}) = \frac{\beta}{6}\int (d\mathbf{p})\frac{d^3 n(\xi_{\mathbf{p}})}{d\xi_{\mathbf{p}}^3} \approx -\frac{\beta}{6}\nu''(0). \quad (9.105)$$

In the last approximation we assumed the temperature (close to ferromagnetic transition) is much smaller than the Fermi energy, so the zero temperature density of states is used. We finally present our expression for the effective Landau-Ginzburg-Wilson (LGW) free energy functional, $F[\mathbf{m}]$, in the vicinity of a ferromagnetic instability, up to the fourth order in the fluctuating magnetization:

$$F[\mathbf{m}] = \frac{1}{\beta}S[\mathbf{m}] \approx \frac{I}{2}\sum_n \int (d\mathbf{q}) \left(1 + \frac{I}{2}\Pi(i\omega_n, \mathbf{q})\right) \mathbf{m}_{i\omega_n,\mathbf{q}} \cdot \mathbf{m}_{-i\omega_n,-\mathbf{q}}$$
$$- \frac{I^4\nu''(0)}{24} \sum_{n,n_1,n_2} \int (d\mathbf{q})(d\mathbf{q}_1)(d\mathbf{q}_2)(\mathbf{m}_{q_1} \cdot \mathbf{m}_{q_2})(\mathbf{m}_{-q_1-q} \cdot \mathbf{m}_{-q_2+q}), \tag{9.106}$$

where $q = (i\omega_n, \mathbf{q})$, $q_1 = (i\omega_{n_1}, \mathbf{q}_1)$, and $q_2 = (i\omega_{n_2}, \mathbf{q}_2)$. For future calculations, it is convenient to use the rescaled fields, $\mathbf{\Psi}_q = \sqrt{I}\mathbf{m}_q$, whereby

$$F[\mathbf{\Psi}] \approx F^{(2)}[\mathbf{\Psi}] + F^{(4)}[\mathbf{\Psi}]$$
$$= \frac{1}{2}\sum_n \int (d\mathbf{q}) \left(1 + \frac{I}{2}\Pi(i\omega_n, \mathbf{q})\right) \mathbf{\Psi}_{i\omega_n,\mathbf{q}} \cdot \mathbf{\Psi}_{-i\omega_n,-\mathbf{q}}$$
$$- \frac{I^2\nu''(0)}{24} \sum_{n,n_1,n_2} \int (d\mathbf{q})(d\mathbf{q}_1)(d\mathbf{q}_2)(\mathbf{\Psi}_{q_1} \cdot \mathbf{\Psi}_{q_2})(\mathbf{\Psi}_{-q_1-q} \cdot \mathbf{\Psi}_{-q_2+q}). \tag{9.107}$$

The negative sign appearing in the quartic term in Eq. (9.107) means that unless $\nu''(0) < 0$, the free energy will run away towards $-\infty$ as $|\mathbf{\Psi}|$ becomes large, making the model unstable. For $\nu''(0) > 0$, one needs to consider higher order terms in the expansion of the free energy until the overall sign is positive, which ensures that the model is stable. Our earlier mean-field treatment made in Sec. 9.5 shows that instead of $\nu''(0) < 0$, we actually need (see Eq. (9.62))

$$\left(\frac{\nu''(0)}{\nu(0)}\right) - 3\left(\frac{\nu'(0)}{\nu(0)}\right)^2 < 0, \tag{9.108}$$

and as we saw, the correction arises from the shift of the chemical potential due to the presence of $\mathbf{m}$. This latter effect is absent in our present path integral analysis since we have dropped the charge degree of freedom. For free electrons in three dimensions, $\nu(0) = A\mu^{1/2}$ and $\nu''(0) < 0$, which means it suffices to expand the free electron functional to quartic order. Also in transition metals where the Fermi level lies near the middle of the narrow incompletely filled d or f bands (i.e. near the peak of the density of states), the condition $\nu''(0) < 0$ is satisfied. Even if the Fermi level is not close to the middle of the narrow band, the steep change in the density of states with respect to energy makes it easy to satisfy the stability condition of Eq. (9.108) since $\nu'(0)$ is large except at an extremum of the density of states.

In the quadratic term in the effective free energy functional, we have the temperature dependent Lindhard polarizability, $\Pi(i\omega_n, \mathbf{q})$. The origin of the temperature dependence is through the Fermi-Dirac distribution function, which means that the characteristic temperature scale is of the order of the Fermi energy. We have of course encountered this in our mean-field analysis of the Stoner instability in Sec. 9.5, where we effectively found that the coefficient of the quadratic term in the free energy has a T^2 temperature dependence[22]. Likewise the temperature dependence of the coefficient of the quartic term in the LGW functional can also be shown to be T^2 with the same characteristic energy scale. We will see below that far more important finite temperature effects arise as a result of low-energy and long-wavelength fluctuations of the order parameter field $\mathbf{\Psi}$, and therefore, we will no longer retain these quadratic dependences in our following analysis.

In our effective model, Eq. (9.107), the dynamics of the order parameter, or spin fluctuation, field $\mathbf{\Psi}$ is entirely generated by the Lindhard polarizability $\Pi(i\omega_n, \mathbf{q}) = (2/\beta)\sum_n \int (d\mathbf{p}) G_0(i\nu_n, \mathbf{p}) G_0(i\nu_n + i\omega_m, \mathbf{p}+\mathbf{q})$. We treat the quadratic term of the effective free energy as the "bare" part and the quartic term as the "interaction". In this Gaussian approximation, the bare correlator of the magnetization fields with four-momenta $q = (i\omega_m, \mathbf{q})$ and $q' = (i\omega'_m, \mathbf{q}')$ is

$$\langle \Psi_{q\alpha} \Psi_{q'\beta} \rangle_0 = \frac{(2\pi)^3}{\beta} \delta_{\alpha\beta} \delta_{m,-m'} \delta(\mathbf{q}+\mathbf{q}') D_0(q), \qquad D_0(q) = \frac{1}{1+\frac{I}{2}\Pi(q)}. \tag{9.109}$$

As $q \to 0$, which is the case we are interested in, the polarizability is determined by the product of two Green functions with nearly coinciding poles. This results in the classic ω/k singularities discussed earlier in Sec. 9.2, where the singular part is dependent on the order in which the limits $\omega \to 0$ and $\mathbf{q} \to 0$ are taken. Comparing with Eq. (9.44), the general expression for the susceptibility, we identify the factor I in Eq. (9.109) as the leading term in the perturbative expansion of the vertex function in the ω-limit, $\Gamma^\omega(q)$, and $\Pi(q)$ is essentially Φ. Recall that $\Gamma^\omega(p, p'; p+k, p'-k)$ consists of all diagrams that do not have any singularity in the limit $k \to 0$. Higher order corrections to the vertex Γ^ω are generated by the quartic term in our path-integral treatment. We can thus expect the path integral and diagram techniques to give identical results. From Eq. (9.44), we also identify our "bare" susceptibility[23],

$$\chi^{(0)}_{\alpha\beta}(i\omega_m, \mathbf{q}) = -\delta_{\alpha\beta}\mu_B^2 \frac{\Pi(q)}{1+\frac{I}{2}\Pi(q)}. \tag{9.110}$$

Consider the Lindhard polarizability at finite temperature and (imaginary) frequency:

$$\Pi(q) = 2\int (d\mathbf{p}) \frac{n(\xi_{\mathbf{p}+\mathbf{q}}) - n(\xi_{\mathbf{p}})}{-i\omega_m + \xi_{\mathbf{p}+\mathbf{q}} - \xi_{\mathbf{p}}}. \tag{9.111}$$

The temperature dependence coming from the distribution functions in the

---

[22] The quadratic temperature dependence, and that too with a characteristic scale of the order of the Fermi energy, contradicts experimental reality, and was one of the reasons we have been looking at fluctuation effects.

[23] Bare, in the sense of having approximated Γ^ω by the bare vertex I.

numerator has a characteristic scale of the order of the Fermi energy and we have already argued that close to the ferromagnetic transition temperature (typically several orders of magnitude less than the Fermi energy), it is not essential to keep this kind of temperature dependence. The more important source of temperature dependence appears in the denominator, through the Matsubara frequencies. If we take the limit $\mathbf{q} \to 0$ before $\omega_m \to 0$, we see that $\Pi(q)$ vanishes. Thus for small q, we need to work with the limit $\mathbf{q} \to 0$ and $\omega_m/|\mathbf{q}| \to 0$, i.e., the so-called $\mathbf{q}$-limit in Landau theory. In this regime, the Lindhard polarizability may be obtained by expanding the numerator in a power-series in $(\xi_{\mathbf{p}+\mathbf{q}} - \xi_{\mathbf{p}})$. We first expand the numerator to leading order in $(\xi_{\mathbf{p}+\mathbf{q}} - \xi_{\mathbf{p}})$ and perform the integration over $\mathbf{p}$. In three dimensions, the result is

$$\Pi(i\omega_m, \mathbf{q}) \approx -\nu(0)\left[1 - \frac{i\omega_m}{2v_F|\mathbf{q}|}\ln\left|\frac{i\omega_m + |\mathbf{q}|v_F}{i\omega_m - |\mathbf{q}|v_F}\right| - \frac{\pi|\omega_m|}{2v_F|\mathbf{q}|}\right], \tag{9.112}$$

which may be obtained by making the analytic continuation $\omega \to i\omega_m$ in Eq. (7.3).

The spin fluctuations are susceptible to decay into a low-energy particle-hole excitation in this regime where $\omega_m/|\mathbf{q}| < 1$.[24] This is an example of Landau damping. The damping of the spin fluctuation modes appears in the form of the non-analytic term, proportional to $|\omega_m|$, in Eq. (9.112). Keeping contributions only to leading order in the small quantity $|\omega_m|/|\mathbf{q}|$, one obtains

$$\Pi(i\omega_m, \mathbf{q}) \approx \frac{\pi\nu(0)}{2v_F}\frac{|\omega_m|}{|\mathbf{q}|}. \tag{9.113}$$

Apart from the above $|\omega_m|/|\mathbf{q}|$ contribution, $\Pi(q)$ also contains terms that depend on $\mathbf{q}$ but not on ω_m. These are not present in Eq. (9.112) (or Eq. (7.3) earlier) as there we had expanded the Fermi distribution function in the numerator only to the first order in $(\xi_{\mathbf{p}+\mathbf{q}} - \xi_{\mathbf{p}})$. The second order contribution at $\omega_m = 0$ is straightforward to calculate. Combining this with Eq. (9.114) we finally obtain

$$\Pi(i\omega_m, \mathbf{q}) \approx -\nu(0)\left(1 - \frac{1}{3}\left(\frac{|\mathbf{q}|}{2k_F}\right)^2 - \frac{\pi}{2v_F}\frac{|\omega_m|}{|\mathbf{q}|}\right). \tag{9.114}$$

We use Eq. (9.114) in the expression for the correlation function of the magnetization, $\langle \Psi_{q\alpha}\Psi_{q'\beta}\rangle$, shown in Eq. 9.109, which now reads as

$$\begin{aligned}\langle \Psi_{q\alpha}\Psi_{q'\beta}\rangle_0 &= \frac{1}{\beta}\frac{(2\pi)^3\delta_{\alpha\beta}\delta_{m,-m'}\delta(\mathbf{q}+\mathbf{q}')}{\left(1 - \frac{I\nu(0)}{2}\right) + \frac{I\nu(0)}{24k_F^2}\mathbf{q}^2 + \frac{\pi I\nu(0)}{4v_F}\frac{|\omega_m|}{|\mathbf{q}|}} \\ &\equiv \frac{1}{\beta}(2\pi)^3\delta_{\alpha\beta}\delta_{m,-m'}\delta(\mathbf{q}+\mathbf{q}')D_0(q). \end{aligned} \tag{9.115}$$

[24] In contrast, the plasma modes we encountered in Chapter 7 are undamped because their energies appear above the particle-hole continuum.

In the second of Eqs. (9.115) we have introduced

$$D_0(q) = \frac{1}{\delta + K\mathbf{q}^2 + \frac{|\omega_m|}{\gamma|\mathbf{q}|}}, \tag{9.116}$$

with the parameters δ, K and γ defined as

$$\delta = 1 - \frac{I\nu(0)}{2} \tag{9.117}$$

$$K = \frac{I\nu(0)}{24k_F^2} \tag{9.118}$$

$$\frac{1}{\gamma} = \frac{\pi I\nu(0)}{4v_F}, \tag{9.119}$$

respectively, have the physical meanings of "distance" from the critical point, spin-wave stiffness and spin-wave damping coefficient. In Eq. (9.115), the momenta $\mathbf{q}$ have a natural upper cutoff, $\Lambda \sim k_F$, at the Fermi momentum, and the frequencies ω_m are to be restricted to a maximum value of $\gamma|\mathbf{q}|$. The latter condition is necessary, for, at higher frequencies, the fluctuations of the order parameter field no longer remain undamped.

Although quantum mechanics adds an extra dimension, i.e., frequency, to the partition function, it is clear from Eq. (9.115) that the frequency and wavevector enter in very different ways. So the quantum critical behavior in d spatial dimensions cannot be inferred from the classical critical behavior in $d+1$ dimensions.

9.6.2 Spin fluctuation effects

In Sec. 9.6.1.2 we obtained an effective LGW free energy functional describing the behavior of the metallic ferromagnet near the critical point. The partition function is

$$\mathcal{Z} = \mathcal{N} \int D[\mathbf{\Psi}] \exp\left[-\frac{\beta}{2}\sum_n \int (d\mathbf{q}) D_0^{-1}(q) \mathbf{\Psi}_{i\omega_n,\mathbf{q}} \cdot \mathbf{\Psi}_{-i\omega_n,-\mathbf{q}} \right.$$
$$\left. -\frac{u\beta}{4} \sum_{n,n_1,n_2} \int (d\mathbf{q})(d\mathbf{q}_1)(d\mathbf{q}_2)(\mathbf{\Psi}_{q_1} \cdot \mathbf{\Psi}_{q_2})(\mathbf{\Psi}_{-q_1-q} \cdot \mathbf{\Psi}_{-q_2+q})\right], \tag{9.120}$$

with $D_0(q)$ defined in Eq. (9.116) and $u = -I^2\nu''(0)/24 > 0$. The parameters in the model are δ, u, and temperature $T = 1/\beta$. The cutoffs on the momentum and energy are Λ and $\gamma|\mathbf{q}|$ respectively. $\mathbf{\Psi}$ is the order parameter or spin fluctuation field. The interaction vertex for the order parameter fields can also be expressed in a fully symmetrized form. Suppose in the interaction term the components $(1, 2, 3)$ of the order parameter vector fields carry the labels $\alpha, \beta, \gamma, \delta$. Diagrammatically we can then denote the bare vertex by, say, an open circle with four legs representing the labels on the order parameter fields, and associate a factor $(u/3)(\delta_{\alpha\beta}\delta_{\gamma\delta}+\delta_{\alpha\gamma}\delta_{\beta\delta}+\delta_{\alpha\delta}\delta_{\beta\gamma})$. The sum of the momenta on the four legs is zero. This fully symmetrized form reflects the fact that in the vertex, the components of the order parameter are pairwise identical.

Critical properties of this model were studied using the mean-field approximation in Sec. 9.5 that ignored the effects of the spatial and temporal fluctuations of the order parameter. Below we will find that the mean-field treatment of Stoner ferromagnetism is inadequate for understanding the correct critical behavior, and it is important to take into account the effect of low-energy and long-wavelength fluctuations of the order parameter field. It turns out that in the quantum case, the upper critical dimension above which the mean-field approximation is justified is not $d = 4$ which is the case in classical LGW models (or even $d = 3$ if the frequencies enter the LGW functional in the same manner as the momenta) but $d - z$, where z, the dynamic exponent, depends on the way in which frequency enters the LGW functional.

9.6.2.1 Perturbative renormalization group

The ideas of scaling theory, developed in the context of classical critical phenomena, can be generalized to the quantum case. In essence, classical scaling theory assumes that near the critical point, the correlation length ξ (or equivalently the distance δ from the critical point) is the only important scale. It then makes sense to integrate out all short length scale fluctuations of the order parameter field up to ξ, resulting in an effective model where the short length scale is now ξ (considerably exceeding the original short length scale that is or the order of some lattice spacing), and the order parameter fields at length scales greater than ξ are only weakly correlated. The procedure of progressively eliminating short length scale fluctuations and keeping track of the effect of such elimination on the parameters of the model is known as the renormalization group (RG). A theoretical scheme[25,26] developed in 1969 and 1971 for understanding critical phenomena combines the renormalization group ideas with perturbation theory. The perturbative RG scheme is analogous to summing certain classes of most divergent diagrams to arbitrary order in perturbation theory in an elegant manner. Hertz in 1976 applied these ideas to the metallic ferromagnets. We will largely follow Hertz's perturbative RG treatment for the quantum critical properties of a metallic ferromagnet. Some errors in Hertz's original treatment for finite temperatures were corrected[27] later and these will also be taken into account in our analysis.

The RG consists of three basic steps – coarse-graining, rescaling and renormalization – that we describe below :

• **Coarse-grain**: In the first step, one separates the field $\mathbf{\Psi}(q)$ into "fast" ($\mathbf{\Psi}_f(q)$) and "slow" ($\mathbf{\Psi}_s(q)$) degrees of freedom where the momenta $|\mathbf{q}|$ lie in

[25]A. I. Larkin and D. E. Khmel'nitskii, *Sov. Phys. JETP* **29**, 1123 (1969).
[26]K. G. Wilson, *Phys. Rev. B* **4**, 3174 (1971); ibid. *Phys. Rev. B* **4**, 3184 (1971).
[27]A. J. Millis, *Phys. Rev. B* **48**, 7183 (1993).

the "shell" $\Lambda > |\mathbf{q}| > \Lambda e^{-d\ell}$ for $\mathbf{\Psi}_f(q)$, and $|\mathbf{q}| < \Lambda e^{-d\ell}$ for $\mathbf{\Psi}_s(q)$. Here $d\ell$ is a small positive number, so we may approximate $e^{d\ell} \approx 1 + d\ell$ where needed. At the Gaussian order, the different momentum modes in the LGW functional are decoupled. On the other hand, in the interaction term, the slow and fast degrees of freedom do not completely separate and one also gets terms that involve coupling of these two kinds of fields (we denote these by $F[\mathbf{\Psi}_f, \mathbf{\Psi}_s]$). Thus

$$\mathcal{Z} = \mathcal{Z}_f \mathcal{N} \int D[\mathbf{\Psi}_s] e^{-\beta F[\mathbf{\Psi}_s]} \left\{ \frac{1}{\mathcal{Z}_f} \int D[\mathbf{\Psi}_f] e^{-\beta F[\mathbf{\Psi}_f]} e^{-\beta F[\mathbf{\Psi}_f, \mathbf{\Psi}_s]} \right\}. \quad (9.121)$$

One now performs the integration over the fast degrees of freedom, which effectively amounts to averaging $e^{-\beta F[\mathbf{\Psi}_f, \mathbf{\Psi}_s]}$ over the fast fields. Note that this integration involves summing over all Matsubara frequencies for these fields. We end up with a new partition function involving only the slow degrees of freedom, now subjected to a momentum cutoff $|\mathbf{q}| < \Lambda e^{-d\ell}$.

• **Rescale**: In the new partition function, there are three types of terms in the LGW functional. First, even in the absence of interactions, we have the Gaussian part. Second, there is an $O(u)$ interaction term involving only the slow degrees of freedom. Let us call this the bare interaction. The third part arises from averaging over the fast degrees of freedom all terms in the LGW functional that involve coupling between the slow and fast fields. This results in corrections to δ, that measures distance from the critical point, as well as the bare interaction u. In the Gaussian part of the full LGW functional, let us rescale the momenta $\mathbf{q} \to \mathbf{q}' = e^{d\ell}\mathbf{q}$ in order to restore the momentum cutoff to the original value Λ. After momentum rescaling the Gaussian part of our LGW functional becomes

$$F^{(2)}[\mathbf{\Psi}_s] = \frac{1}{2} \sum_n \int^{\Lambda} (d\mathbf{q}') e^{-3d\ell} \left(\delta + K e^{-2d\ell} |\mathbf{q}'|^2 + e^{d\ell} \frac{|\omega_n|}{\gamma |\mathbf{q}'|} \right) |\mathbf{\Psi}_s(i\omega_n, \mathbf{q}')|^2. \quad (9.122)$$

• **Renormalize**: In the final step of the RG, we renormalize (a) the fields $\mathbf{\Psi}_s$, to keep the coefficient of the $\mathbf{q}^2$ term fixed at K, and (b) rescale the Matsubara frequencies (or temperature) to keep the coefficient of the $|\omega_n|/|\mathbf{q}|$ term fixed at $1/\gamma$. This is of course not a unique choice, and is guided by physical considerations. The physical reason for this choice is that at the critical point (i.e. $\delta = 0$) the model should be scale-free, which in turn requires that the above coefficients do not change upon scaling. The new temperature $T' = T(\ell + d\ell)$ is related to the old one through $T(\ell + d\ell) = e^{z\ell} T(\ell) = e^{3d\ell} T(\ell)$, or,

$$\frac{dT(\ell)}{d\ell} = zT(\ell). \quad (9.123)$$

We have introduced the dynamic exponent z which in our case has the value $z = 3$. A value of z different from unity means that the model shows different scaling with respect to momenta and frequencies. We need to find the scaling relation between the renormalized fields $\mathbf{\Psi}$ and the old ones $\mathbf{\Psi}_s$. For this, we consider the action for the slow fields given by $\beta F[\mathbf{\Psi}_s]$, and noting that we have rescaled the temperature above, we rewrite the action as $\beta' e^{-z d\ell} F[\mathbf{\Psi}_s]$. If we now define the renormalized fields by $\mathbf{\Psi} = e^{-((3+z+2)/2)d\ell}\mathbf{\Psi}_s$, that would absorb the $e^{-3d\ell}$ arising from rescaling of $(d\mathbf{q})$, $e^{-2d\ell}$ coming from the rescaling of the $\mathbf{q}^2$ term, and the factor $e^{-zd\ell}$ from rescaling the temperature. Rescaling the momenta and temperature also changes the (Matsubara) frequency cutoff $\gamma|\mathbf{q}|$. This cutoff continuously varies with $\gamma|\mathbf{q}|$, while the Matsubara frequencies change only in discrete steps. Consequently, at high enough temperatures where the distance between successive Matsubara frequencies is large, it is not clear how this cutoff should be imposed. Hertz's ad-hoc prescription was to treat the frequencies as a continuum at low temperatures but keep only the zero frequency term at high temperatures[28]. The above rescaling and renormalizations also result in rescaling of the interaction u, the quantity δ, and, as is clear from Eq. (9.121), even the normalization factor $\mathcal{N}$. Following the RG procedure, Hertz obtained scaling equations for δ and the interaction u.

A better way for treating the issue of cutoffs at finite temperature was suggested by Millis[29] who expressed the Matsubara sum over imaginary frequencies by an integral over real frequencies. The cutoff and rescaling are done with respect to these real frequencies, and the procedure is correct even at finite temperatures. We will follow the latter strategy. A key ingredient in Millis' approach consists of first integrating over all fluctuations of the order parameter (without separating into slow and fast degrees of freedom). Subsequently, the separation into contributions from slow and fast degrees of freedom is done through the usual coarse-graining and rescaling process.

It is instructive to first consider the Gaussian, noninteracting model. We perform the Gaussian integration over the order parameter fields and write the partition function as $\mathcal{Z} = e^{-\beta\Omega^{(0)}}$, with $\Omega^{(0)}$ the thermodynamic potential for the non-interacting model:

$$\Omega^{(0)} = \frac{nV}{2\beta}\sum_m \int (d\mathbf{q}) \ln\left(\delta + K\mathbf{q}^2 + \frac{|\omega_m|}{\gamma|\mathbf{q}|}\right). \tag{9.124}$$

Here $n = 3$ is the number of components of the (vector) order parameter field. Observe that $\delta + K\mathbf{q}^2 > 0$ as we are on the paramagnetic side of the transition. We analytically continue the frequencies to the complex domain (with Matsubara frequencies chosen to lie along the imaginary axis). In the upper half plane, $|\omega_m|$ analytically continues to $-i\omega$, while in the lower half-plane, $|\omega_m|$ continues to $i\omega$. Clearly, the logarithmic function, which in the upper half-plane is of the form $\ln(A - iB\omega)$, with A, $B > 0$, does not have a singularity anywhere in the upper half-plane. Likewise the logarithm does not have any

[28]Such an approximation does not give the correct description of the finite temperature regime

[29]See footnote 27.

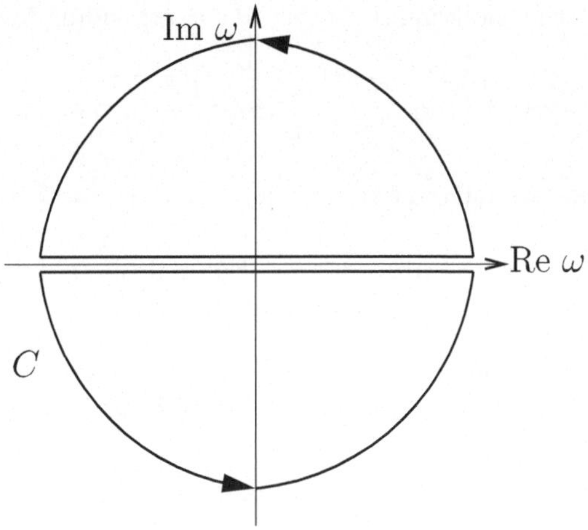

FIGURE 9.7: Contour for performing the Matsubara summation in Eq. (9.124). The summand is singular along the line $\text{Im}\omega = 0$. The bosonic Matsubara frequencies lie along the $\text{Im}\omega$ axis. The Matsubara sum is equivalent to the sum of two line integrals along the real axis (see Eq. (9.126)).

singularity in the lower half plane. The only singularity of the logarithmic term is on the line $\text{Im}\omega = 0$. To perform the sum over Matsubara frequencies, we can therefore consider contour integrals of the form $\oint_C (d\omega) \coth(\beta\omega/2) W(\omega)$, with C as shown in Fig. 9.7[30].

The function

$$W(\omega) = \begin{cases} \ln(A - iB\omega), & \text{Im}\omega > 0, \\ \ln(A + iB\omega), & \text{Im}\omega < 0. \end{cases} \tag{9.125}$$

The integral vanishes over the big semi-circles. Then, using the residue theorem,

$$\begin{aligned} \frac{1}{2\pi i} \oint_C d\omega \, \coth(\beta\omega/2) W(\omega) &= \frac{1}{2\pi i} \int_{-\infty}^{\infty} d\omega \, \coth(\beta\omega/2) \ln\left(\frac{A - iB\omega}{A + iB\omega}\right) \\ &= \frac{2}{\beta} \sum_m \ln(A + B|\omega_m|). \end{aligned} \tag{9.126}$$

The logarithmic term in Eq. (9.126) may be cast in a more familiar form using $\ln[(A - iB\omega)/(A + iB\omega)] = -2i \tan^{-1}(B\omega/A)$. Collecting Eq. (9.124) and Eq.

[30]The function $\coth(\beta\omega/2)$ has simple poles at the bosonic Matsubara frequencies with residue $2/\beta$.

(9.126), the thermodynamic potential corresponding to the Gaussian term is[31]

$$\Omega^{(0)} = -nV \int^{\Lambda} (d\mathbf{q}) \int_0^{\gamma|\mathbf{q}|} \frac{d\omega}{\pi} \coth\left(\frac{\beta\omega}{2}\right) \tan^{-1}\left(\frac{\omega/\gamma|\mathbf{q}|}{\delta + K\mathbf{q}^2}\right). \tag{9.127}$$

Observe that we have introduced appropriate cutoffs on the momentum and frequency.

We apply our RG procedure to Eq. (9.127). The first (coarse-grain) step consists of elimination of momenta in the shell $\Lambda > |\mathbf{q}| > \Lambda e^{-d\ell}$, which yields

$$\begin{aligned}\Omega^{(0)} \approx -nV &\int^{\Lambda e^{-d\ell}} (d\mathbf{q}) \int_0^{\gamma|\mathbf{q}|} \frac{d\omega}{\pi} \coth\left(\frac{\beta\omega}{2}\right) \tan^{-1}\left(\frac{\omega/\gamma|\mathbf{q}|}{\delta + K\mathbf{q}^2}\right) \\ &- d\ell\, nV\Lambda^3 \mathcal{K}_3 \int_0^{\gamma\Lambda} \frac{d\omega}{\pi} \coth\left(\frac{\beta\omega}{2}\right) \tan^{-1}\left(\frac{\omega/\gamma\Lambda}{\delta + K\Lambda^2}\right),\end{aligned} \tag{9.128}$$

with $\mathcal{K}_3 = S_3/(2\pi)^3$, S_3 being the surface area of a three dimensional sphere of unit radius. Next we rescale the momenta to $\mathbf{q}' = e^{d\ell}\mathbf{q}$, which requires us to rescale the frequency to $\omega' = e^{zd\ell}\omega$, the temperature to $T' = Te^{zd\ell}$, the parameter δ to $\delta' = e^{2d\ell}\delta$, and the volume to $V' = e^{-3d\ell}V$. We re-express Eq. (9.128) in terms of the new variables and parameters. In Eq. (9.128), the integration over ω extends to $\gamma|\mathbf{q}|$, which means the upper limit on the ω' integration is[32] $\gamma|\mathbf{q}|e^{zd\ell} = \gamma|\mathbf{q}'|e^{2d\ell}$. We eliminate frequencies ω' in the shell $\gamma|\mathbf{q}'|e^{2d\ell} > \omega' > \gamma|\mathbf{q}'|$ so that the new cutoff for ω' is $\gamma|\mathbf{q}'|$. At this stage the thermodynamic potential can be written in a form that explicitly brings out the changes upon rescaling:

$$\begin{aligned}\beta\Omega^{(0)} \approx -nV'\beta' &\int^{\Lambda} (d\mathbf{q}') \int_0^{\gamma|\mathbf{q}'|} \frac{d\omega'}{\pi} \coth\left(\frac{\beta'\omega'}{2}\right) \tan^{-1}\left(\frac{\omega'/\gamma|\mathbf{q}'|}{\delta' + K\mathbf{q}'^2}\right) \\ &- d\ell\, nV\beta\Lambda^3 \mathcal{K}_3 \int_0^{\gamma\Lambda} \frac{d\omega}{\pi} \coth\left(\frac{\beta\omega}{2}\right) \tan^{-1}\left(\frac{\omega/\gamma\Lambda}{\delta + K\Lambda^2}\right) \\ &- d\ell\, \frac{2nV\beta\gamma}{\pi} \int^{\Lambda} (d\mathbf{q}')|\mathbf{q}'| \coth\left(\frac{\beta'\gamma|\mathbf{q}'|}{2}\right) \tan^{-1}\left(\frac{1}{\delta' + K\mathbf{q}'^2}\right) \qquad (9.129) \\ &\equiv (\beta'\Omega'^{(0)}) - d\ell\, n f^{(0)}. \qquad (9.130)\end{aligned}$$

[31]Note that $\Omega^{(0)}$ has the opposite sign to Millis' result.

[32]We have used $(z-1) = 2$ here because $\omega/\gamma_{\mathbf{q}}$ always scales as $\mathbf{q}^2$ by design, regardless of the actual form of the damping $\gamma_{\mathbf{q}}$. Indeed z is determined by this condition. For a clean metallic ferromagnet, which is what we are analyzing here, $\gamma_{\mathbf{q}} = \gamma|\mathbf{q}|$. For a metallic antiferromagnet, $\gamma_{\mathbf{q}} = \text{const}$.

Here we have introduced the scale-dependent quantity,

$$f^{(0)}(T(\ell),\delta(\ell)) = V(\ell)\beta(\ell) \times \left[\Lambda^3 K_3 \int_0^{\gamma\Lambda} \frac{d\omega}{\pi} \coth\left(\frac{\beta(\ell)\omega}{2}\right) \tan^{-1}\left(\frac{\omega/\gamma\Lambda}{\delta(\ell)+K\Lambda^2}\right) + \frac{2\gamma}{\pi} \int^{\Lambda} (d\mathbf{q})|\mathbf{q}| \coth\left(\frac{\beta(\ell)\gamma|\mathbf{q}|}{2}\right) \tan^{-1}\left(\frac{1}{\delta(\ell)+K\mathbf{q}^2}\right)\right], \tag{9.131}$$

for future use. We identify the first term on the RHS of Eq. (9.130) as $\beta'\times$ the free energy of the slow degrees of freedom. The remaining terms amount to $-d\ell n f^{(0)}(T)$, and we identify it as the contribution from the integrated out fast degrees of freedom. Clearly, $\mathcal{Z}_f(\ell+d\ell) = \mathcal{Z}_f(\ell)e^{d\ell n f^{(0)}(T(\ell),\delta(\ell))}$, i.e.,

$$\frac{d\mathcal{Z}_f(\ell)}{d\ell} = n f^{(0)}(T(\ell),\delta(\ell))\mathcal{Z}_f(\ell). \tag{9.132}$$

The initial condition is $\mathcal{Z}_f(0) = 1$, for at the beginning of the RG procedure, no fast modes have been eliminated yet. By integrating the above scaling equation, we obtain $\ln(\mathcal{Z}_f(\ell))$, which is needed for calculating certain thermodynamic properties such as the specific heat (see Sec. 9.6.2.3). Since the sign of $f^{(0)}$ is positive, we infer that $\mathcal{Z}_f$ increases as slow degrees of freedom are progressively integrated out. We may also interpret $d\ell\, n f^{(0)}$ in Eq. (9.130) as the change in the thermodynamic potential, $d\Omega^{(0)}$, upon infinitesimal coarse graining and rescaling,

$$d\Omega^{(0)}(\ell) = \frac{\partial\Omega^{(0)}(\ell)}{\partial\ell} d\ell = +nT(\ell)f^{(0)}(\ell)d\ell. \tag{9.133}$$

Next we consider the effect of the interactions on the thermodynamic potential. We treat the interaction term as a perturbation and perform the average over the order parameter fields, $\langle e^{-S^{(4)}[\mathbf{\Psi}]}\rangle_{\mathbf{\Psi}}$, with $S^{(4)}$ the quartic term in Eq. (9.120). The thermodynamic potential assumes the form of an expansion in cumulants of $S^{(4)}[\mathbf{\Psi}]$:

$$\begin{aligned}\Omega &= \Omega^{(0)} - \frac{1}{\beta}\ln[\langle e^{-S^{(4)}[\mathbf{\Psi}]}\rangle] \\ &= \Omega^{(0)} + \frac{1}{\beta}\langle S^{(4)}[\mathbf{\Psi}]\rangle - \frac{1}{2\beta}\left(\langle S^{(4)}[\mathbf{\Psi}]^2\rangle - \langle S^{(4)}[\mathbf{\Psi}]\rangle^2\right) + \ldots \\ &= \Omega^{(0)} + \Omega^{(1)} + \Omega^{(2)} + \ldots.\end{aligned} \tag{9.134}$$

The first term in the cumulant expansion of the interaction contains averages of the type $\sum_{\alpha\beta}\langle\Psi_\alpha\Psi_\alpha\Psi_\beta\Psi_\beta\rangle$, where α, β run over the n components of our vector order parameter field, $\mathbf{\Psi}$. There are $n(n+2)$ ways of performing pairwise averages; hence, $\sum_{\alpha\beta}\langle\Psi_\alpha\Psi_\alpha\Psi_\beta\Psi_\beta\rangle = n(n+2)\langle\Psi\Psi\rangle\langle\Psi\Psi\rangle$, where the two-point averages are as shown in Eq. (9.115). The contribution of the first cumulant to the thermodynamic potential is

$$\Omega^{(1)} = \frac{n(n+2)}{4} u \sum_{n,n_1,n_2} \int (d\mathbf{q})(d\mathbf{q}_1)(d\mathbf{q}_2) \langle \Psi_{q_1} \Psi_{q_2} \rangle \langle \Psi_{-q_1-q} \Psi_{-q_2+q} \rangle$$
$$= Vu \frac{n(n+2)}{4\beta^2} \sum_{n,n_1} \int (d\mathbf{q})(d\mathbf{q}_1) D_0(i\omega_{n_1}, \mathbf{q}_1) D_0(i\omega_n, \mathbf{q}). \qquad (9.135)$$

The factor $n(n+2)$ may also be obtained by considering the fully symmetrized structure for the interaction vertex, $(1/3)(\delta_{\alpha\beta}\delta_{\gamma\delta} + \delta_{\alpha\gamma}\delta_{\beta\delta} + \delta_{\alpha\delta}\delta_{\beta\gamma})$, and setting pairs of indices equal and tracing. For example, setting $\alpha = \beta$, $\gamma = \delta$ and tracing gives $(1/3)(n^2 + 2n)$. We now sum over the Matsubara frequencies following the same procedure that we followed for $\Omega^{(0)}$. To evaluate $(1/\beta)\sum_n D_0(i\omega_n, \mathbf{q})$, we analytically continue the Matsubara frequencies to the complex plane and express the sum in the form of a contour integral making use of the residue theorem:

$$\frac{2}{\beta} \sum_n \frac{1}{\delta + K\mathbf{q}^2 + \frac{|\omega_n|}{\gamma|\mathbf{q}|}} = \frac{1}{2\pi i} \oint_C d\omega \coth\left(\frac{\beta\omega}{2}\right) \times \begin{cases} \frac{1}{\delta + K\mathbf{q}^2 - i(\omega/\gamma|\mathbf{q}|)}, & \mathrm{Im}\omega > 0 \\ \frac{1}{\delta + K\mathbf{q}^2 + i(\omega/\gamma|\mathbf{q}|)}, & \mathrm{Im}\omega < 0 \end{cases}, \qquad (9.136)$$

and C is the contour shown in Fig. 9.7. As usual, the integral vanishes over the big semi-circles leaving us with

$$\frac{1}{\beta} \sum_n \frac{1}{\delta + K\mathbf{q}^2 + \frac{|\omega_n|}{\gamma|\mathbf{q}|}} = \int_{-\infty}^{\infty} (d\omega) \coth\left(\frac{\beta\omega}{2}\right) \mathrm{Im}\left[\frac{1}{\delta + K\mathbf{q}^2 - i(\omega/\gamma|\mathbf{q}|)}\right], \qquad (9.137)$$

which, upon substitution in Eq. (9.135) leads us to our final expression for $\Omega^{(1)}$:

$$\Omega^{(1)} = Vu \frac{n(n+2)}{4} \left[\int^{\Lambda} (d\mathbf{q}) \int_0^{\gamma|\mathbf{q}|} \frac{d\omega}{\pi} \coth\left(\frac{\beta\omega}{2}\right) \times \frac{(\omega/\gamma|\mathbf{q}|)}{(\delta + K\mathbf{q}^2)^2 + (\omega/\gamma|\mathbf{q}|)^2} \right]^2$$
$$\equiv Vu \frac{n(n+2)}{4} [J^{(1)}(\Lambda, \delta, T)]^2, \qquad (9.138)$$

with

$$J^{(1)}(\Lambda, \delta, T) = \int^{\Lambda} (d\mathbf{q}) \int_0^{\gamma|\mathbf{q}|} \frac{d\omega}{\pi} \coth\left(\frac{\beta\omega}{2}\right) \frac{(\omega/\gamma|\mathbf{q}|)}{(\delta + K\mathbf{q}^2)^2 + (\omega/\gamma|\mathbf{q}|)^2}. \qquad (9.139)$$

It is easily seen from Eq. (9.127) that the function $J^{(1)}$ is just the partial derivative of $\Omega^{(0)}$ with respect to δ :

$$\frac{\partial \Omega^{(0)}}{\partial \delta} = nVJ^{(1)}. \tag{9.140}$$

This relation enables us to calculate the shift in δ due to the interactions.

Next we consider the second order interaction correction in the cumulant expansion for the thermodynamic potential in Eq. (9.134), i.e., $\Omega^{(2)} = -(1/2\beta)\langle (S^{(4)})^2\rangle_c$, with the subscript c there to remind us that only connected diagrams need to be taken into account. Diagrammatically, the second order correction to the vertex is obtained by contracting any two legs from one of the bare vertices with two from the second[33]. That leaves four legs free, two from each bare vertex. We attach the labels $\alpha, \beta, \gamma, \delta$ to these four legs, with the Greek indices denoting the component of the order parameter field. As usual, the independent internal degrees of freedom are to be summed over. Summing over the vector components in the internal lines gives us a factor $(1/9)(n+8)(\delta_{\alpha\beta}\delta_{\gamma\delta} + \text{permutations})$ for the fully symmetrized vertex. Since there are ${}^4C_2 = 6$ ways of selecting a pair each of external legs on the bare vertices, we have an additional factor of 36. The most singular contribution of the second order correction to the vertex happens when the pair of internal D_0 lines transfers zero net four-momentum. Such a situation corresponds to forward scattering of the spin fluctuation modes. These diagrams are obtained by joining the free ends at either end of the ladder so that there is no net momentum transferred by the internal pair of lines. Each loop at the ends provides a factor n, and we have an additional factor of n^2 in this manner. Collecting all these results, the second order contribution to the free energy is

$$\begin{aligned}\Omega^{(2)} &= -u^2 V \frac{n^2(n+8)}{8\beta^3} \sum_{\{n\}} \int \prod_{i=1}^{3} (d\mathbf{q}_i) D_0(i\omega_{n_1}, \mathbf{q}_1) \\ &\qquad \times [D_0(i\omega_{n_2}, \mathbf{q}_2)]^2 D_0(i\omega_{n_3}, \mathbf{q}_3) \\ &\equiv -u^2 V \frac{n^2(n+8)}{4} [J^{(1)}(\Lambda, \delta, T)]^2 J^{(2)}(\Lambda, \delta, T),\end{aligned} \tag{9.141}$$

where the function $J^{(1)}(\Lambda, \delta, T)$ is as defined in Eq. (9.139) and $J^{(2)}(\Lambda, \delta, T)$ is given by

$$J^{(2)}(\Lambda, \delta, T) = \int^{\Lambda} (d\mathbf{q}) \int_0^{\gamma|\mathbf{q}|} \frac{d\omega}{\pi} \coth\left(\frac{\beta\omega}{2}\right) \frac{(\omega/\gamma|\mathbf{q}|)(\delta + K\mathbf{q}^2)}{[(\delta + K\mathbf{q}^2)^2 + (\omega/\gamma|\mathbf{q}|)^2]^2}. \tag{9.142}$$

The second line of Eq. (9.141) is obtained as usual by analytic continuation to complex frequencies and expressing the Matsubara sums as contour integrals

[33] In other words, a ladder diagram with two vertices.

over the frequencies. Observe that the second order contribution to the free energy has an opposite sign to the first order contribution. Indeed, it can be seen that successive terms in the perturbative expansion of the thermodynamic potential come with alternating signs. The functions $J^{(1)}$ and $J^{(2)}$ are related through

$$\frac{\partial^2 \Omega^{(0)}}{\partial \delta^2} = nV \frac{\partial J^{(1)}}{\partial \delta} = -2nVJ^{(2)}. \tag{9.143}$$

We may now apply our RG procedure to the perturbative corrections to the free energy, $\Omega^{(1)}$ and $\Omega^{(2)}$ exactly as we did for $\Omega^{(0)}$ earlier. However it is more convenient to first look at the effect of coarse graining and rescaling on the functions $J^{(1)}$ and $J^{(2)}$. Consider first the function $J^{(1)}$. Upon coarse graining we obtain

$$J^{(1)}(\Lambda, \delta, T) \approx J^{(1)}(\Lambda e^{-d\ell}, \delta, T) + d\ell\, \mathcal{K}_3 \Lambda^3 \times \int_0^{\gamma\Lambda} \frac{d\omega}{\pi} \coth(\beta\omega/2) \frac{(\omega/\gamma\Lambda)}{(\delta + K\Lambda^2)^2 + (\omega/\gamma\Lambda)^2}. \tag{9.144}$$

Next we rescale the momenta in the expression for $J^{(1)}(\Lambda e^{-d\ell}, \delta, T)$ to restore the cutoff at Λ, which, as before, requires us to simultaneously rescale the temperature ($T' = e^{zd\ell}T$), frequencies ($\omega' = e^{zd\ell}\omega$) and the parameter δ to $\delta' = e^{2d\ell}\delta$. Since this process changes the frequency cutoff for ω' to $\gamma|\mathbf{q}'|e^{2d\ell}$, we eliminate modes in the frequency range $\gamma|\mathbf{q}'| < \omega' < \gamma|\mathbf{q}'|e^{2d\ell}$ in order to restore the frequency cutoff to $\gamma|\mathbf{q}'|$. After the rescaling step,

$$\begin{aligned} J^{(1)}(\Lambda, \delta, T) &\approx e^{-(z+d-2)d\ell} J^{(1)}(\Lambda, \delta', T') + d\ell\, \mathcal{K}_3 \Lambda^3 \\ &\quad \times \int_0^{\gamma\Lambda} \frac{d\omega}{\pi} \coth(\beta\omega/2) \frac{(\omega/\gamma\Lambda)}{(\delta + K\Lambda^2)^2 + (\omega/\gamma\Lambda)^2} \\ &\quad + d\ell \frac{2\gamma}{\pi} \int^{\Lambda} (d\mathbf{q}')|\mathbf{q}'| \coth(\beta'\gamma|\mathbf{q}'|/2) \frac{1}{(\delta' + K\mathbf{q}'^2)^2 + 1}. \end{aligned} \tag{9.145}$$

As in Eq. (9.131), we introduce another scale-dependent quantity,

$$\begin{aligned} f^{(2)}(T(\ell), \delta(\ell)) &= \mathcal{K}_3 \Lambda^3 \int_0^{\gamma\Lambda} \frac{d\omega}{\pi} \coth(\beta(\ell)\omega/2) \frac{(\omega/\gamma\Lambda)}{(\delta(\ell) + K\Lambda^2)^2 + (\omega/\gamma\Lambda)^2} \\ &\quad + \frac{2\gamma}{\pi} \int^{\Lambda} (d\mathbf{q})|\mathbf{q}| \coth(\beta(\ell)\gamma|\mathbf{q}|/2) \frac{1}{(\delta(\ell) + K\mathbf{q}^2)^2 + 1}. \end{aligned} \tag{9.146}$$

The first order correction to the free energy, $\Omega^{(1)}$, now reads (see Eq. (9.138))

$$\begin{aligned}\beta\Omega^{(1)} &\approx (V\beta)ue^{-2(z+d-2)d\ell}\frac{n(n+2)}{4}J^{(1)}(\Lambda,\delta',T')\\&\quad\times\left[J^{(1)}(\Lambda,\delta',T')+2d\ell e^{(z+d-2)d\ell}f^{(2)}\right]\\&\approx (V'\beta')u'\frac{n(n+2)}{4}[J^{(1)}(\Lambda,\delta',T')]^2\\&\quad+d\ell(V\beta)u\frac{n(n+2)}{2}J^{(1)}(\Lambda,\delta,T)f^{(2)}\\&\equiv(\beta'\Omega'^{(1)})+d\ell\beta u\frac{(n+2)}{2}\frac{\partial\Omega^{(0)}}{\partial\delta}f^{(2)}.\end{aligned}\tag{9.147}$$

Here we used the relation in Eq. (9.140) between $J^{(1)}$ and $\partial\Omega^{(0)}\partial\delta$, and also introduced the scaling of the interaction, $u\to u'=e^{-(z-1)d\ell}u\equiv e^{-(d+z-4)d\ell}u$.

By taking the partial derivative of $f^{(0)}$ with respect to $\delta(\ell)$ in Eq. (9.131) and comparing with Eq. (9.146), we obtain

$$\frac{\partial f^{(0)}}{\partial\delta(\ell)}=-V(\ell)\beta(\ell)f^{(2)},\tag{9.148}$$

which is similar to Eq. (9.140) that related the function $J^{(1)}$ to the partial derivative of $\Omega^{(0)}$ with respect to δ.

We repeat the same coarse-graining and rescaling procedure for the function $J^{(2)}$, which yields

$$J^{(2)}(\Lambda,\delta,T)\approx e^{(d+z-4)d\ell}J^{(2)}(\Lambda,\delta',T')+d\ell f^{(4)},\text{ with}\tag{9.149}$$

$$\begin{aligned}f^{(4)}(T(\ell),\delta(\ell)) &= K_3\Lambda^3\int_0^{\gamma\Lambda}\frac{d\omega}{\pi}\coth(\beta(\ell)\omega/2)\frac{(\omega/\gamma\Lambda)(\delta+K\Lambda^2)}{[(\delta+K\Lambda^2)^2+(\omega/\gamma\Lambda)^2]^2}\\&\quad+\frac{2\gamma}{\pi}\int^{\Lambda}(d\mathbf{q})|\mathbf{q}|\coth(\beta(\ell)\gamma|\mathbf{q}|/2)\frac{\delta+K\mathbf{q}^2}{[(\delta+K\mathbf{q}^2)^2+1]^2}.\end{aligned}\tag{9.150}$$

The function $f^{(4)}$ is related to the derivatives of $f^{(0)}$ and $f^{(2)}$ with respect to δ :

$$\frac{\partial^2 f^{(0)}}{\partial\delta(\ell)^2}=-V(\ell)\beta(\ell)\frac{\partial f^{(2)}}{\partial\delta(\ell)}=2V(\ell)\beta(\ell)f^{(4)}.\tag{9.151}$$

We use Eqs. (9.149) and (9.145) in Eq. (9.141) to obtain the behavior of the second correction to the thermodynamic potential $\Omega^{(2)}$ upon performing the RG procedure:

$$\begin{aligned}\beta\Omega^{(2)} &\approx -(u')^2(V'\beta')\frac{n^2(n+8)}{4}[J^{(1)}(\Lambda,\delta',T')]^2J^{(2)}(\Lambda,\delta',T')\\&\quad-d\ell\,u^2(V\beta)\frac{n^2(n+8)}{4}[2J^{(1)}J^{(2)}f^{(2)}+(J^{(1)})^2f^{(4)}].\end{aligned}\tag{9.152}$$

We are now in a position to calculate the perturbative corrections to the scaling equations for δ and u. Consider the last line of Eq. (9.147). We identify $\beta u\frac{(n+2)}{2}f^{(2)}$ as $\partial\delta/\partial\ell$, as that allows us to interpret the term $d\ell\beta u\frac{(n+2)}{2}(\partial\Omega^{(0)}/\partial\delta)f^{(2)}$ as $(\partial\Omega^{(0)}/\partial\delta)(\partial\delta/\partial\ell)d\ell$, i.e., the free energy of eliminated modes due to changes in scaling of δ on account of the interaction. Taking this correction into account together with the trivial scaling $\delta' = \delta e^{2d\ell}$ yields

$$\frac{d\delta}{d\ell} = 2\delta(\ell) + \frac{1}{2}(n+2)f^{(2)}(T(\ell), \delta(\ell))u(\ell). \tag{9.153}$$

We can alternatively read off the shift in δ to leading order in u,

$$\delta(u,\ell) \approx \delta(0,\ell) + \frac{1}{2}(n+2)J^{(1)}(\Lambda, \delta(0,\ell), T(\ell))u(\ell), \tag{9.154}$$

and perform the RG process with $\delta(u,\ell)$. In the context of Eq. (9.154), we would like to mention a self-consistent renormalization (SCR) scheme originally developed by Moriya, Ramakrishnan and Lonzarich[34,35,36]. In the SCR scheme, the interacting theory is approximated by a Gaussian model with a renormalized δ, whose value is self-consistently determined by replacing $\delta(0,\ell)$ on the RHS of Eq. (9.154) with $\delta(u,\ell)$. This approximation accounts for many of the experimental features of metallic ferromagnets that are not accounted for by the Stoner mean-field theory (such as the temperature dependence of the susceptibility in the quantum critical region). It also agrees with the Hertz-Millis theory in giving correct exponents for critical scaling in the regime where the properties are essentially those of a Gaussian model. However the crossover to this regime is not correctly described because of the inherent limitations of the approximation.

Similarly considering the scaling of $\Omega^{(2)}$ in Eq. (9.150), we may read off the perturbative correction to the scaling equation for u. Combining with the trivial rescaling $u' = ue^{-(d+z-4)d\ell}$, we obtain

$$\frac{du}{d\ell} = -(d+z-4)u(\ell) - \frac{1}{4}(n+8)f^{(4)}(T(\ell), \delta(\ell))[u(\ell)]^2. \tag{9.155}$$

In summary, we have four RG equations, namely, (9.153) and (9.155) for the parameters δ and u respectively, together with (9.132) for the partition function and (9.123) for the temperature.

We need to solve the above four RG equations to understand the effect of spin fluctuations on the physical properties of the itinerant ferromagnet. Before proceeding to do that, some basic terminology needs to be introduced. A very important notion in the RG description of a quantum system is that

[34] See T. Moriya, *Spin Fluctuations in Itinerant Electron Magnets* (Springer-Verlag, Berlin, 1985).

[35] T.V. Ramakrishnan, *Phys. Rev. B* **10**, 4014 (1974).

[36] G. G. Lonzarich and L. Taillefer, *J. Phys. C* **18**, 4339 (1985).

of a *fixed point* – particular values of the parameters δ, T, u governing the free energy functional that do not change under the RG scaling procedure. Since under coarse-graining, δ must rescale, a fixed point corresponds to $\delta = 0$ or $\delta = \infty$. An obvious fixed point for our RG equations is $\delta = T = u = 0$. It is an *unstable* fixed point because for small deviations of the initial values of the parameters from the fixed point, at least some of the parameters, here δ and T, grow upon rescaling, taking us away from the fixed point. On the other hand, near a *stable* fixed point, the parameters flow towards the fixed point upon rescaling. The stability of the fixed point is determined by considering the flow of different parameters in the free energy functional upon rescaling, with initial values of the parameters chosen in the vicinity of the fixed point. In addition to the parameters in the free energy functional, one may also ask how an arbitrary quantity evolves upon rescaling. This leads us to the concept of relevant, irrelevant and marginal operators. To determine the relevance of a certain quantity (operator), one examines the linearized RG equations in the vicinity of the fixed point. Quantities that move away from the fixed point upon rescaling are called *relevant* operators while those that converge towards the fixed point are called *irrelevant* operators. Quantities that do not change upon rescaling the linearized RG equations are termed as *marginal*, and one needs to consider higher order (nonlinear) terms in the RG equations to determine their relevance.

For future calculations, it is convenient to redefine the momenta and energies so that Λ, and γ are set to unity. We also set $V = 1$ and the stiffness $K = 1$ without any loss of generality.

9.6.2.2 Solution of Hertz-Millis RG equations

We will approach the critical point from the paramagnetic side, i.e. $\delta > 0$. The RG equations have an unstable fixed point at $\delta = T = u = 0$, and we will be looking at solutions in the vicinity of this fixed point. For our three-dimensional ferromagnet, the quantity $d + z - 4 = 2$ is positive. That means for small initial values of the interaction, the scaling of u is initially toward $u = 0$ (i.e., u is an irrelevant operator). In this regime, let us therefore ignore the u^2 correction in the scaling equation for u, and consider the simpler RG equations for the parameters,

$$\frac{dT}{d\ell} = zT(\ell), \tag{9.156}$$

$$\frac{d\delta}{d\ell} = 2\delta(\ell) + \frac{1}{2}(n+2)f^{(2)}(T(\ell))u(\ell), \tag{9.157}$$

$$\frac{du}{d\ell} = -(d+z-4)u(\ell). \tag{9.158}$$

We have dropped the dependence of $f^{(2)}$ on δ as it is not important for our subsequent analysis. The above fixed point evidently corresponds to the Gaussian or noninteracting limit. The RG equations are readily solved (for small positive initial values δ, T, u):

$$T(\ell) = T\,e^{z\ell}, \tag{9.159}$$

$$u(\ell) = u\,e^{-(d+z-4)\ell}, \tag{9.160}$$

$$\delta(\ell) = e^{2\ell}\left(\delta + \frac{u}{2}(n+2)\int_0^\ell dx\,e^{-(d+z-2)x} f^{(2)}(T\,e^{zx})\right). \tag{9.161}$$

We stop the scaling once $\delta(\ell) \sim 1$ is reached (with the original cutoffs, scaling is to be stopped at $\delta(\ell) \sim K\Lambda^2$).

In our solution for $\delta(\ell)$, we have an integral over $f^{(2)}$, which we evaluate in different physical regimes. At the point scaling is stopped, the value of the temperature $T(\ell)$ can be large or small compared to the frequency cutoff $\gamma\Lambda$ that we have set to unity for convenience.

If $T(\ell) \ll 1$ when scaling is stopped, then we should make a low-temperature expansion for $f^{(2)}$ and integrate term by term. This is the so-called quantum regime in Hertz theory. To find the extent of the quantum regime, we set $\delta = 1$ and $f^{(2)}(T(x)) \approx f^{(2)}(T = 0) = \frac{1}{2}(\ln 2)\mathcal{K}_3$ in Eq. (9.161).

Performing the integration, we obtain the scale ℓ_q at which $\delta(\ell_\mathrm{q}) = 1$ at low temperatures,

$$\ell_\mathrm{q} \approx -\frac{1}{2}\ln\left(\delta + \frac{u(n+2)f^{(2)}(0)}{2(d+z-2)}\right), \tag{9.162}$$

which translates to the temperature scale

$$T_\mathrm{q} = Te^{z\ell_\mathrm{q}} \equiv \frac{T}{r^{z/2}}, \text{ with}$$

$$r = \delta + \frac{u(n+2)f^{(2)}(0)}{2(d+z-2)}. \tag{9.163}$$

For $T(\ell_\mathrm{q}) < 1$, i.e., $T < r^{z/2}$, we are in the quantum regime[37] where the properties are governed by the Gaussian fixed point at $T = 0$. Note that the interactions cause a shift in the quantum critical point from $\delta = 0$ (the Stoner value) to $r = 0$.

If $T(\ell) > 1$ when the scaling is stopped, then we need to proceed differently. We calculate δ in Eq. (9.161) in two steps. First, we calculate δ for the temperature $T(\ell_1) = 1$, i.e.,

$$\ell_1 = \frac{1}{z}\ln(1/T). \tag{9.164}$$

The temperature dependence of $f^{(2)}$ may no longer be dropped. In the low

[37]With the original cutoff, the quantum regime corresponds to $T < \gamma\Lambda(r/K\Lambda^2)^{z/2}$.

temperature regime, $T(x) < 1$, we separate $f^{(2)}$ into a zero temperature part and a finite temperature correction. Thus we approximate

$$f^{(2)}(T(x)) \approx \begin{cases} f^{(2)}(0) + [f^{(2)}(T(x)) - f^{(2)}(0)], & T(x) < 1 \\ cT(x), & T(x) > 1, \end{cases} \tag{9.165}$$

where

$$c = \frac{\mathcal{K}_3}{2} + \frac{4\mathcal{K}_3}{\pi} \int_0^1 dq \frac{q^2}{1+q^4} = \frac{\mathcal{K}_3}{2} + \frac{\mathcal{K}_3}{\sqrt{2}\pi}[\pi + \ln(3 - 2\sqrt{2})]. \tag{9.166}$$

We now use Eq. (9.165) in Eq. (9.161). At $T(\ell_1) = 1$, the value of $\delta(\ell_1) = \delta_1$ is

$$\begin{aligned} \delta_1 &= T^{-2/z}\left(r + \frac{Bu}{2}(n+2)T^{(d+z-2)/z}\right) \\ &\equiv T_{\mathrm{q}}^{-2/z}\left(1 + \frac{Bu}{2r}(n+2)T^{(d+z-2)/z}\right), \end{aligned} \tag{9.167}$$

with

$$B \approx -\frac{f^{(2)}(0)}{d+z-2} + \int_0^1 (f^{(2)}(y) - f^{(2)}(0))y^{-(d+2z-2)/z}. \tag{9.168}$$

The lower limit of the integral in Eq. (9.168) is in fact T but in this low temperature regime, we have set it at zero since that involves small errors of order T. With this, the quantity B becomes independent of the temperature. The interaction also scales to the value $u(\ell_1) = u_1$ at $T(\ell_1) = 1$:

$$u_1 = u\, e^{-(d+z-4)\ell_1} = u\, T^{(d+z-4)/z} < u. \tag{9.169}$$

Next we extend our scaling to the region $T(\ell) > 1$, where $f^{(2)} \approx cT$. The RG begins at the scale ℓ_1 with initial parameter values δ_1 and u_1 as obtained in Eq. (9.167) and Eq. (9.169) respectively. The RG is stopped at $\ell = \ell_{\mathrm{cl}}$ such that $\delta(\ell_{\mathrm{cl}}) = 1$. We use $f^{(2)}(T(x)) \approx cT(x)$ in the integral in Eq. (9.161),

$$\int_{\ell_1}^{\ell_{\mathrm{cl}}} e^{-(d+z-2)x} f^{(2)}(T\, e^{zx}) \approx \frac{cT}{d-2}(e^{-(d-2)\ell_1} - e^{-(d-2)\ell_{\mathrm{cl}}}), \tag{9.170}$$

whence

$$1 = e^{2\ell_{\mathrm{cl}}}\left(r + \frac{B_1 u}{2}(n+2)T^{(d+z-2)/z} - cT\frac{u(n+2)}{2(d-2)}e^{-(d-2)\ell_{\mathrm{cl}}}\right), \tag{9.171}$$

$$\text{and } B_1 = B + \frac{c}{d-2}. \tag{9.172}$$

The approximate solution for $\ell_{\rm cl}$ is[38]

$$\ell_{\rm cl} \approx -\frac{1}{2}\ln\left(r + \frac{B_1 u}{2}(n+2)T^{(d+z-2)/z}\right). \tag{9.173}$$

There is a better way that brings out the physics of the $T(\ell) > 1$ region in a more elegant manner, and is completely equivalent to the solutions we have obtained above. We note that Eq.(9.158) for δ contains the combination $f^{(2)}u$. Since $f^{(2)}(T) \approx cT$ in the region under consideration, we rewrite the RG equations for δ and also the interaction u in terms of a new parameter, $v(\ell) = u(\ell)T(\ell)$:

$$\begin{aligned}\frac{d\delta}{d\ell} &= 2\delta(\ell) + \frac{1}{2}(n+2)c\,v(\ell)\\ \frac{dv}{d\ell} &= -(d-4)v(\ell) - \frac{1}{4}(n+8)g\,\delta(\ell)[v(\ell)]^2.\end{aligned} \tag{9.174}$$

The scaling in Eq. (9.174) commences at $\ell_1 = \frac{1}{z}\ln(1/T)$ with initial values δ_1 and $v(\ell_1) = u_1 T(\ell_1) = u_1$, and runs up to a scale approximately given by $\ell_{\rm cl}$. These equations are essentially identical to a classical n-component ferromagnet with interaction parameter v. At ℓ_1, it is easily verified that the quantities δ_1 and u_1 are both much less than unity, so we drop the nonlinear in v contribution in the second of Eq. (9.174). Observe that for $d < 4$, the parameter v initially increases upon scaling. In RG theory, this is an example of a relevant perturbation. Above $d = 4$, the interaction is irrelevant and the critical properties of the model revert to a Gaussian mean-field theory. Formally, the dimension above which interactions become irrelevant is known as the *upper critical dimension.*

In the regime $T(\ell) > 1$, our effective interaction parameter $v(\ell)$ *increases* upon scaling in the manner

$$v(\ell) = u_1\, e^{(4-d)(\ell-\ell_1)}. \tag{9.175}$$

The solution for δ is also easily found:

$$\begin{aligned}\delta(\ell) &= e^{2\ell}\left(e^{-2\ell_1}\delta_1 + \frac{cu_1(n+2)}{2}e^{-(4-d)\ell_1}\frac{e^{-(d-2)\ell_1} - e^{-(d-2)\ell}}{(d-2)}\right)\\ &= e^{2(\ell-\ell_1)}\left(\delta_1 + \frac{cu_1(n+2)}{2}\frac{1-e^{-(d-2)(\ell-\ell_1)}}{(d-2)}\right).\end{aligned} \tag{9.176}$$

Both $v(\ell)$ and $\delta(\ell)$ are functions of $\bar{\ell} = (\ell - \ell_1)$. Our calculations are based on the premise that the interaction v remains small when scaling is stopped at $\delta(\ell_{\rm cl}) = 1$. For $d = 3$, we can see from Eq. (9.175) that this condition is tantamount to $u_1 e^{(\ell_{\rm cl}-\ell_1)} \ll 1$. The solution for $e^{(\ell_{\rm cl}-\ell_1)}$ obtained from the condition $\delta(\ell_{\rm cl}) = 1$ in Eq. (9.176) is $e^{(\ell_{\rm cl}-\ell_1)} \approx (\delta_1 + cu_1(n+2)/2)^{-1/2}$. This

[38] Compare this with $\ell_{\rm q} = -\frac{1}{2}\ln(r)$ for the quantum regime.

gives us the following condition for ignoring higher order corrections to the interaction:

$$\frac{u_1}{\sqrt{\delta_1 + cu_1(n+2)/2}} \ll 1, \text{ or,}$$
$$\frac{uT}{\sqrt{r + T^{(1+z)/z}B_1 u(n+2)/2}} \ll 1 \tag{9.177}$$

If $r < 0$, the condition fails at some characteristic temperature T_c where the denominator becomes small and comparable to the small quantity in the numerator, uT_c. This gives us an estimate for the transition temperature:

$$T_c \simeq \left(\frac{2|r|}{B_1 u(n+2)}\right)^{z/(z+1)}. \tag{9.178}$$

We also identify Eq. (9.177) as the Ginzburg criterion for ignoring fluctuation effects and using the mean-field approximation. For the range of parameters where $T > r^{z/2}$ and Eq. (9.177) are both satisfied, the critical behavior of our metallic ferromagnet is governed by a (classical) d-dimensional Gaussian fixed point. For $T < r^{z/2}$ we have already seen that the critical behavior is that of a (quantum) $d + z$-dimensional Gaussian theory. In the classical region, if the initial value of the interaction v is large enough such that $\delta(\ell) < 1$ at the point where $v(\ell)$ has already reached unity, our linearized RG equations are inadequate and we must look at higher order or perhaps non-perturbative corrections.

9.6.2.3 Physical consequences of spin fluctuations

The solutions of the Hertz-Millis RG equations are useful in understanding physical consequences of the spin fluctuations. We shall discuss the behavior of the correlation function ξ and magnetic susceptibility, and also the specific heat coefficient $\gamma = C_V/T$. Here we will come across instances where spin fluctuation effects lead to the breakdown of the Landau Fermi liquid in three dimensions.

Let us recall the rationale for our perturbative RG treatment. As order parameter fluctuations are statistically self-similar at all length scales up to the correlation length ξ, one may progressively eliminate the short length scale fluctuations until one ends up with a model with a coarse-grained characteristic length scale ξ and corresponding characteristic energies $\Delta \sim \xi^{-z}$. Note that the characteristic energy scale evolves as $\Delta \sim \xi^{-z}$ for all temperatures. This is a consequence of scaling theory – to keep the spin-wave stiffness K and damping coefficient γ fixed[39], the frequency rescaling must go as $\omega \to e^{zd\ell}\omega$ for wavevector rescaling $\mathbf{q} \to \mathbf{q}e^{d\ell}$. We consider the above physical quantities in the different regimes discussed in Sec. 9.6.2.1.

[39] Required on physical grounds, that at the critical point, the model should be scale-free, see Sec. 9.6.2.1.

• *Quantum regime,* $T < r^{z/2}$: The scale $\ell_q = -\frac{1}{2}\ln(r)$ at which scaling is stopped determines the correlation length, i.e., $\xi \simeq e^{\ell_q} = r^{-1/2}$. The critical exponent $\nu = 1/2$ for the correlation length, $\xi \simeq |r|^{-\nu}$, is in accordance with the Gaussian fixed point behavior. The corresponding characteristic energy is $\Delta \simeq r^{z/2}$. The susceptibility has the dependence $\chi \simeq 1/r$ for Gaussian criticality.

To estimate the specific heat coefficient, consider Eq. (9.132) that describes the scaling of $\mathcal{Z}_f$, the partition function corresponding to the RG eliminated degrees of freedom. Integrating Eq. (9.132) from $\ell = 0$ to $\ell = \ell_q = -\frac{1}{2}\ln(r)$ yields

$$\begin{aligned}
\ln \mathcal{Z}_f(\ell_q) &= n\int_0^{\ell_q} dx\, V(x)\beta(x) f^{(0)}(T\, e^{zx}) \\
&= nV\beta \int_0^{\ell_q} dx\, e^{-(d+z)x}\left(f^{(0)}(0) + T(x)\frac{df^{(0)}}{dT}\bigg|_{T=0} + \dots\right) \\
&= nV\beta \int_0^{\ell_q} dx\, e^{-(d+z)x}\left(f^{(0)}(0) + C_2 T(x)^2 + \dots\right). \qquad (9.179)
\end{aligned}$$

Here, the constant

$$C_2 = (8\mathcal{K}_3/\pi)\int_0^{\infty} dy\, y^2/\sinh^2 y. \qquad (9.180)$$

The thermodynamic potential of the eliminated degrees of freedom is $-T\ln \mathcal{Z}_f$. The contribution proportional to $f^{(0)}$ gives us the temperature-independent part of the thermodynamic potential of the eliminated modes. The remaining part gives

$$\begin{aligned}
-T\ln(\mathcal{Z}_f) &= -nVC_2T^2 \int_0^{\ell_q} dx\, e^{-(d+z)x} e^{2zx} \\
&= -nVC_2T^2\ell_q = -\frac{1}{2}nVC_2T^2\ln(1/r). \qquad (9.181)
\end{aligned}$$

The thermodynamic potential corresponding to the slow degrees of freedom, Ω_s, is just[40] $e^{-z\ell_q}\Omega^{(0)}(\ell_q)$:

$$\Omega_s(\ell_q) = -e^{-(d+z)\ell_q} nV\left[\int_0^1 (d\mathbf{q})\int_0^{|\mathbf{q}|}\frac{d\omega}{\pi}\coth\left(\frac{e^{-z\ell_q}\omega}{2T}\right)\tan^{-1}\left(\frac{\omega/|\mathbf{q}|}{1+\mathbf{q}^2}\right)\right]. \qquad (9.182)$$

We used $\delta(\ell_q) = 1$. The quantity in square brackets is a function of $T_q = T\, e^{z\ell_q}$. At low temperatures, the thermodynamic potential of the slow modes can be shown to have the expansion

$$e^{-z\ell_q}\Omega^{(0)}(\ell_q) \approx -e^{-(d+z)\ell_q} nV\left(C_0 + C_2' T_q^2 - C_4 T_q^4 \ln(1/T_q) + \dots\right), \qquad (9.183)$$

[40]see Eq. (9.130) for an explanation.

with C_0, C_2' and C_4 positive constants. Of these, the second term makes a small correction to the larger $T^2 \ln(1/r)$ contribution from the fast modes as can be seen from Eq. (9.181). The C_4 term vanishes in the limit of zero temperature but nonetheless has a logarithmic singularity, so we keep it. Combining Eq. (9.181) and Eq. (9.183), we obtain our final approximate expression for the thermodynamic potential in the quantum regime:

$$\Omega \approx -\frac{1}{2} nVC_2T^2 \ln(1/r) + nVC_4T^4 \frac{1}{r^3} \ln(r^{z/2}/T). \tag{9.184}$$

The specific heat $C_V \equiv \gamma(T)T$ is $-(T/V)(\partial^2\Omega/\partial T^2)$, from which we obtain the coefficient of specific heat,

$$\gamma(T) \approx nC_2 \ln(1/r) - 12nC_4T^2 \frac{1}{r^3} \ln(r^{z/2}/T). \tag{9.185}$$

Two noteworthy features in Eq. (9.185) must be pointed out. First, at low temperatures, the specific heat coefficient of our quantum critical ferromagnet, $\gamma(T)$, tends to a constant, just as in a Landau Fermi liquid. However there is a significant enhancement by a factor $\ln(1/r)$ that diverges at the quantum critical point, $r = 0$. This we interpret as enhancement of effective quasiparticle mass in Landau theory. Second, there are significant higher order $(1/r^3)T^2 \ln(1/T)$ corrections to $\gamma(T)$ that differ from T^2 contributions (Debye law) that one expects from other low-energy bosonic excitations such as phonons. The difference comes from the peculiar nature of our overdamped spin fluctuations. The characteristic energy scale of these damped spin fluctuations, $\sim r|\mathbf{q}$, can be much smaller (near the quantum critical point) than phonon energies at the same wavevectors, and such damped spin-waves therefore get excited in much larger numbers than phonons. Insofar as the specific heat is concerned, the spin fluctuations in the quantum regime do not invalidate the Landau Fermi liquid picture in three dimensions, but they nonetheless cause significant renormalization of the effective mass. The result in Eq. (9.185) was in fact first obtained in 1966 in the context of paramagnon effects in liquid ^{3}He, long before the development of Hertz theory, by studying the effect of spin fluctuations on the fermion self-energies[41,42]. Such a self-energy diagram is shown in Fig. 9.6.

• *Classical regime, $T > r^{z/2}$*: The classical regime consists of two distinct parts. For temperatures larger than T_q but not too large such that the perturbative correction in Eq. (9.173), $(B_1u/2r)(n+2)T^{(1+z)/z}$, is small, then $\ell_{cl} \approx \ell_q$, and the correlation length is still given by $\xi \simeq r^{-1/2}$ as in the quantum regime. At higher temperatures where $(B_1u/2r)(n+2)T^{(1+z)/z} > 1$, then it is evident from Eq. (9.173) that the scale at which the RG process is stopped

[41] S. Doniach and S. Engelsberg, *Phys. Rev. Lett.* **17**, 750 (1966)

[42] The quadratic in T contribution to $\gamma(T)$ differs from Millis' in having an extra power of r in the denominator but agrees with Doniach and Engelsberg, 1966.

is $\ell_c \simeq -\frac{1}{2}\ln((B_1u/2)(n+2)T^{(1+z)/z})$. This gives us the correlation length,

$$\xi \simeq e^{\ell_{cl}} \simeq 1/\sqrt{(B_1u/2)(n+2)T^{(1+z)/z}} \sim T^{-2/3}, \tag{9.186}$$

for the high temperature part of the classical regime. It is evident from the z-dependence of the correlation length that this regime, named "classical" by Hertz, nevertheless depends crucially on quantum dynamics. Since in both the classical and quantum regimes, the theory is effectively Gaussian, the susceptibility shows the mean-field behavior, $\chi \sim \xi^2$.

To obtain the specific heat coefficient, we calculate the thermodynamic potential. For the fast degrees of freedom one has

$$\Omega_f = -T\ln Z_f = -nV\int_0^{\ell_{cl}} dx\, e^{-(d+z)x} f^{(0)}(Te^{zx}). \tag{9.187}$$

The integration is performed over the two intervals, $0 < x < \ell_1 = (1/z)\ln(1/T)$ and $\ell_1 < x < \ell_{cl}$. For $0 < x < \ell_1$, make a low-temperature expansion, separating $f^{(0)}(T(x))$ respectively into zero temperature and finite temperature contributions, $f^{(0)}(0)$ and $f^{(0)}(T(x)) - f^{(0)}(0) \approx C_2T(x)^2$, with C_2 defined in Eq. (9.180). Integrating the part containing $f^{(0)}(0)$ results in a constant plus a non-singular temperature-dependent part that goes as $e^{-(d+z)\ell_1} = T^{(d+z)/z} \equiv T^2$. These are not important for us. The remaining part gives a singular contribution, $-nVC_2T^2\ell_1 = -\frac{1}{z}nVC_2T^2\ln(1/T)$. Now we consider the interval $\ell_1 < x < \ell_{cl}$, where we use the approximation

$$f^{(0)}(T(x)) \approx C_1T(x),\ T(x) \gg 1. \tag{9.188}$$

The contribution to the thermodynamic potential is

$$-\frac{1}{3}nVC_1T(e^{-3\ell_1} - e^{-3\ell_{cl}}) \simeq -\frac{1}{3}nVC_1T[T - \xi^{-3}],$$

and is not singular. For the slow degrees of freedom, since $T(\ell_{cl}) > 1$, we simplify $\coth(\omega/2T(\ell_{cl})) \approx 2T(\ell_{cl})/\omega$:

$$\begin{aligned}\Omega_s(\ell_{cl}) = e^{-z\ell_{cl}}\Omega^{(0)} &\approx -2nVTe^{-3\ell_{cl}}\int_0^1 (d\mathbf{q})\int_0^{|\mathbf{q}|}\frac{d\omega}{\pi}\frac{1}{\omega}\tan^{-1}\left(\frac{\omega/|\mathbf{q}|}{1+\mathbf{q}^2}\right)\\ &\equiv -nVET\xi^{-3}.\end{aligned} \tag{9.189}$$

Combining the contributions from the fast and slow degrees of freedom, the thermodynamic potential in the classical regime is

$$\Omega \approx -\frac{1}{z}nVC_2T^2\ln(1/T) - nVE_1T\xi^{-3}, \tag{9.190}$$

where $E_1 = E - C_1/3$. The coefficient of specific heat is

$$\gamma(T) \approx \frac{2n}{z}C_2\ln\left(\frac{1}{T}\right). \tag{9.191}$$

The singular $\ln(1/T)$ behavior of $\gamma(T)$ signals the breakdown of Landau Fermi liquid theory in the classical regime of Hertz-Millis theory for a three dimensional metal on the verge of a ferromagnetic transition. The singular behavior is a consequence of the presence of a continuum of low-energy long-wavelength spin fluctuations with energies much less than the temperature: summing over these low-energy modes results in the singularities.

We now discuss the resistivity in the classical Gaussian regime. Recall that in clean metals, the resistivity at low temperatures is dependent on the number density of low-energy phonons; these result in largely forward scattering of the electrons. From Fermi's Golden Rule, the electron-phonon transport scattering rate for small momentum transfer $\mathbf{q}$ is proportional to $|\mathbf{q}|^5$, with $|\mathbf{q}|^3$ coming from phase space factors and the remaining power of two coming from the factor $(1-\cos\theta_{\mathbf{q}}) \sim |\mathbf{q}|^2$ that appears in expressions for the transport scattering rate[43] Because of the linear relation between phonon energies and wavevectors, the $|\mathbf{q}|^5$ dependence translates to the well-known T^5 temperature dependence of resistivity. In Hertz theory, the spin fluctuation energies go as the third power of the wavevector, not linear. Consequently, the resistivity obeys a $T^{5/3}$ law. This law has been confirmed in actual measurements in the vicinity of a ferromagnetic QCP[44].

In Fig. 9.8 we present a schematic phase diagram showing the different physical regimes discussed in this section.

How do spin fluctuations affect the physical properties in nearly *antiferromagnetic* metals? Since the wavevector for antiferromagnetic ordering, $\mathbf{Q}$ is finite, the spin-wave damping term now has the form $|\omega_n|/\gamma_{\mathbf{Q}}$, with $\gamma_{\mathbf{Q}} \sim \gamma|\mathbf{Q}|$ essentially a constant. Upon performing the RG procedure, it is immediately evident that the dynamical exponent $z = 2$. Our results for the critical temperature T_c and the correlation length for magnetic order, ξ, still hold, provided $z = 2$ is used instead of $z = 3$ that was the case for the ferromagnetic metal. The specific heat coefficient may be similarly calculated although it needs a bit more care (see Millis (1993)). Here we simply quote the results. In the quantum regime in three dimensions,

$$\gamma(T) \approx \gamma_0(r) - nC_4'\frac{T^2}{r^{3/2}}, \tag{9.192}$$

where $\gamma_0(r)$ is not singular as $r \to 0$ and C_4' is a constant. In the classical regime,

$$\gamma(T) \approx \gamma_0(r) - E_1'T^{1/z}, \tag{9.193}$$

and E_1' is a constant.

In the following chapters, we will look at numerous other examples of

---

[43] Physically, it means backscattering processes are more effective at momentum relaxation than forward scattering processes.

[44] R. P. Smith, M. Sutherland, G. G. Lonzarich, S. S. Saxena, N. Kimura, S. Takashima, M. Nohara and H. Takagi, *Nature* **455**, 1220 (2008).

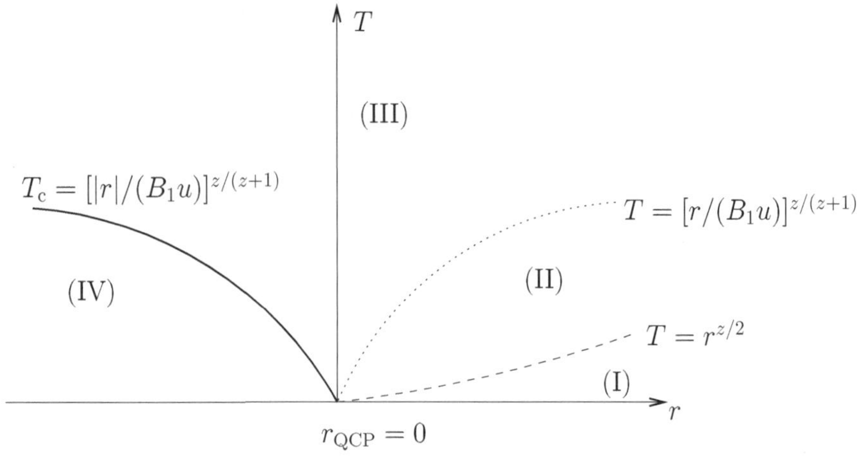

FIGURE 9.8: Schematic T vs r phase diagram showing different physical regimes in the Hertz-Millis theory of a three-dimensional metal near a magnetic quantum critical point (QCP). At the QCP, $r = 0$. The parameter r may be tuned by pressure, doping or some other means. The dynamic exponent $z = 3$ for the ferromagnetic case and $z = 2$ for the antiferromagnetic case. Region (I) is the "quantum" Gaussian regime where Landau Fermi liquid theory holds. This is also the regime of study in Doniach-Engelsberg theory. Here the susceptibility $\chi \sim r^{-1}$ as in mean field theory, and the specific heat coefficient has a singular behavior, $\gamma(T) \sim \ln(1/r)$, for the ferromagnet and a nonsingular constant (in the limit $r \to 0$) for the antiferromagnet. The "classical" regime is subdivided into Regions (II) and (III). Region (II) is the perturbative classical regime where $\chi(T)$ is as in the Fermi liquid regime but the specific heat coefficient for the ferromagnet, $\gamma(T) \sim \ln(1/T)$, is different from the quantum regime. The antiferromagnet shows no divergence but has the form $\gamma(T)-\gamma(0) \sim T^{1/z}$. In Region (III), the finite temperature terms dominate the temperature independent ones. Here $\chi(T) \sim T^{-z/1+z}$ and $\gamma(T) \sim \ln(1/T)$. The curve $T_c = [|r|/(B_1u)]^{z/(z+1)}$ is the boundary for classical phase transition to a magnetically ordered phase (Region (IV)). In the literature, it is common to describe Region (I) as the Fermi liquid regime and Regions (II) and (III) as the "quantum critical" regime, since in the latter case, the critical exponents depend sensitively on the dynamic exponent z that governs quantum effects.

departures from Fermi liquid behavior in metals on account of spin fluctuation effects.

Part III

Beyond Fermi Liquid Theory

Chapter 10

Effect of Ampere Forces in Fermi Liquids

When a metal is subjected to a static but spatially slowly varying external charge density, we found in Sec. 6.1 that the conduction electrons screen this perturbation beyond λ_{TF}, the Thomas-Fermi screening length. Since the electric field $\mathbf{E}$ in this case is purely determined by the Poisson law, $\boldsymbol{\nabla}\cdot\mathbf{E} = -\rho/\epsilon_0$, it is purely longitudinal, or, curl-free. Thus, purely longitudinal electric fields get screened by the conduction electrons. A magnetic field $\mathbf{B}$ is in contrast purely transverse (divergence-free), and a static magnetic field is in fact not screened by the conduction electrons. When the electromagnetic fields are time-dependent, apart from the magnetic field which is always transverse, the electric field may also acquire a transverse component. In Sec. 6.7, it was shown that time-dependent magnetic fields are screened beyond the skin depth $\lambda_{\text{skin}} = 1/\sqrt{\sigma\mu\mu_0\omega}$, with σ the conductivity, μ the magnetic permittivity and μ_0 the permittivity of vacuum. The lack of screening of transverse fluctuations of the electromagnetic fields in a metal indicates in turn strong effects on the electronic properties. Some of these effects will be discussed here.

On a related note, we also point out that in the Hertz theory of nearly ferromagnetic metals (see Sec. 9.6), the spin fluctuation fields also have long-range correlations. The lack of screening of the slowly varying magnetization fluctuations in turn has strong effects on the properties of the system such as the divergence of the specific heat coefficient in the "classical Gaussian regime. Although the spin fluctuations couple to the electrons through their spin degree of freedom but not to their momentum, we shall see there are nonetheless numerous parallels between the two.

10.1 Effective action for transverse electromagnetic fluctuations in metals

To perform the microscopic analysis, it is convenient to work with the scalar and potentials, ϕ and $\mathbf{A}$ respectively, instead of the physical electromagnetic fields $\mathbf{E}$ and $\mathbf{B}$. We choose the gauge condition $\boldsymbol{\nabla}\cdot\mathbf{A}=0$, for in that case the correlation functions of the scalar and vector potentials decouple. To manage notation clutter, we will work with CGS units and also set the speed of light in vacuum, c, to one. After the calculations have been performed, we can always revert to our preferred MKS units. In the absence of particles, the electromagnetic fields have the following action in Euclidean time,

$$S_{\mathrm{em}}[A]=\frac{1}{2}\int d\tau\, d\mathbf{r}\,\left[(\boldsymbol{\nabla}\phi)^2+(\partial_t\mathbf{A})^2+(\boldsymbol{\nabla}\times\mathbf{A})^2\right]. \tag{10.1}$$

The coupling of the gauge fields to nonrelativistic fermions is described by the action,

$$\begin{aligned}S_{\mathrm{el}}[A,c]&=\int dx\, c_\sigma^\dagger\left[\partial_\tau-\epsilon_F+e\phi+\frac{1}{2m}(\mathbf{p}-e\mathbf{A})^2\right]c_\sigma\\ &\equiv\int dx\, c_\sigma^\dagger\left[-G_0^{-1}+e\phi-\frac{e}{m}\mathbf{A}\cdot\mathbf{p}+\frac{e^2}{2m}\mathbf{A}^2\right]c_\sigma,\end{aligned} \tag{10.2}$$

where $x\equiv(\tau,\mathbf{r})$ and A denotes the four vector $(\phi,\mathbf{A})$. We have ignored the coupling of the spin degree of freedom to the magnetic field for the moment. The partition function is

$$\mathcal{Z}=\int D[c,A]e^{-S_{\mathrm{el}}[A,c]-S_{\mathrm{em}}[A]}, \tag{10.3}$$

and we choose the Landau or transverse gauge, $\boldsymbol{\nabla}\cdot\mathbf{A}=0$. Here $G_0^{-1}(i\nu_n,\mathbf{p})=i\nu_n-\mathbf{p}^2/2m+\epsilon_F=i\nu_n-\xi(\mathbf{p})$ is the bare Green function for the conduction electrons and $\nu_n=(2\pi/\beta)(n+1/2)$ is a fermionic Matsubara frequency. With our choice of gauge, correlation functions involving the vector potential must satisfy the following Ward-Takahashi identity,

$$\boldsymbol{\nabla}_{\mathbf{r}}\cdot\langle\mathbf{A}(\mathbf{r},\tau)\times\text{any other fields}\rangle=0. \tag{10.4}$$

Recall that the effective action in our earlier study of Hertz theory was obtained by a Hubbard-Stratonovich transformation that rendered the action bilinear in the fermion fields (see Eq. (9.93)), followed by integration over the fermion fields in Eq. (9.94). The situation is rather similar in Eq. (10.3).

In this chapter, we are interested in Ampere effects, or in other words, the effects of transverse electromagnetic field fluctuations. Since only the scalar potential ϕ is associated with the longitudinal component of the electric field, we will hereafter regard the ϕ fields in the action as integrated out, with the Coulomb interaction effects incorporated into effective Fermi liquid parameters.

The next step is to integrate out the fermionic fields, just as we did for Hertz theory. The resulting action is

$$S[\mathbf{A}] = S_{\mathrm{em}}[\mathbf{A}] - 2\mathrm{Tr}\ln\left[1 - G_0\left(-\frac{e}{m}\mathbf{A}\cdot\mathbf{p} + \frac{e^2}{2m}\mathbf{A}^2\right)\right]. \tag{10.5}$$

The factor of two is on account of spin degeneracy and the trace is over all remaining degrees of freedom. Now we expand the logarithm in powers of $\mathbf{A}$ treating the charge e as a small parameter:

$$\begin{aligned} S[\mathbf{A}] \approx \frac{1}{2}\int d\tau\, d\mathbf{r}\, \left[(\partial_\tau \mathbf{A})^2 + (\boldsymbol{\nabla}\times\mathbf{A})^2\right] \\ + \frac{e^2}{m}\mathrm{Tr}\left[G_0\mathbf{A}^2 + \frac{1}{m}G_0\mathbf{A}\cdot\mathbf{p}G_0\mathbf{A}\cdot\mathbf{p}\right] + O(\mathbf{A}^4). \end{aligned} \tag{10.6}$$

Let us simplify the terms on the RHS of Eq. (10.6) a bit. We choose to work in the momentum space:

$$\mathbf{A}_{\mathbf{r}}(\tau) = \sum_{n=-\infty}^{\infty}\int(d\mathbf{q})\mathbf{A}(q)e^{-i\omega_n\tau}. \tag{10.7}$$

ω_n are bosonic Matsubara frequencies respectively, and $q = (i\omega_n, \mathbf{q})$. In the momentum representation, the effective action takes the following form:

$$\begin{aligned} S[\mathbf{A}] \approx{}& \frac{\beta}{2}\sum_{s;a}\int(d\mathbf{q})A_a(i\omega_s,\mathbf{q})A_a(-i\omega_s,-\mathbf{q})\left(\omega_s^2 + \mathbf{q}^2\right) \\ &+ \frac{e^2\beta}{2m}\sum_{s;a,b}\int(d\mathbf{q})A_a(i\omega_s,\mathbf{q})A_b(-i\omega_s,-\mathbf{q}) \\ &\times\left[n\delta_{ab} + \frac{2}{\beta m}\sum_k\int(d\mathbf{p})G_0(i\nu_k,\mathbf{p}^{(-)})G_0(i\nu_k + i\omega_s,\mathbf{p}^{(+)})p_a^{(-)}p_b^{(+)}\right]. \end{aligned} \tag{10.8}$$

Here $n = 2G_0(x = 0) \equiv \frac{2}{\beta}\sum_n\int(d\mathbf{k})G_0(i\nu_n,\mathbf{k})$ is the conduction electron density and $\mathbf{p}^{(\pm)} = \mathbf{p} \pm \frac{1}{2}\mathbf{q}$. In the absence of the coupling to conduction electrons, the two-point correlation function of the vector potentials is of the form $M_{ab}/(\omega_s^2 + \mathbf{q}^2)$, where M is a 3×3 matrix projecting to the transverse space, satisfying $M^2 = M$ and $\sum_a q_a M_{ab} = 0$. This is required in order to satisfy the Ward-Takahashi identity imposed by our gauge choice. It is easily seen that

$$M_{ab} \propto \delta_{ab} - \frac{q_a q_b}{\mathbf{q}^2}. \tag{10.9}$$

With the coupling to the fermions taken into account, the two-point correlator is still proportional to M_{ab} on account of the Ward-Takahashi identity.

To proceed further we need to evaluate the quantity in square brackets in Eq. (10.8). The quantity in square brackets in Eq. (10.8) is related to the electromagnetic response function[1],

$$K_{ab}(i\omega_s, \mathbf{q}) = \frac{e^2}{m|\omega_s|}\left[n\delta_{ab} + \frac{2}{\beta m}\sum_k \int (d\mathbf{p}) G_0(i\nu_k, \mathbf{p}^{(-)}) \right.$$
$$\left. \times G_0(i\nu_k + i\omega_s, \mathbf{p}^{(+)}) p_a^{(-)} p_b^{(+)}\right]. \tag{10.10}$$

The two terms are referred to in the literature as the diamagnetic and paramagnetic components of the electromagnetic response function. Taking the limit $|\mathbf{q}| \to 0$, $\omega_s/|\mathbf{q}| \to 0$ (the $\mathbf{q}$-limit) leads to vanishing of the quantity in square brackets on account of exact cancellation of its two terms. To see this, we perform the Matsubara sum in the second term and take the above limit, whereupon it gives us

$$-\frac{2}{m}\int (d\mathbf{p}) p_a\, p_b \frac{dn}{d\xi(\mathbf{p})} = -\delta_{ab}\frac{2}{3}\nu(0)\int d\xi\,(\xi + \epsilon_F)\frac{dn}{d\xi} = -\delta_{ab} n, \tag{10.11}$$

completely canceling the diamagnetic part. Here we used $\langle p_a p_n\rangle = \frac{1}{3}p^2\delta_{ab} = \frac{2m}{3}(\xi + \epsilon_F)$ which is the case for isotropic systems. On the other hand, taking the ω-limit, $\omega_s \to 0$, $|\mathbf{q}|/\omega_s \to 0$ kills off the second term.

Consider now Eq. (10.8) and Eq. (10.10). We assume without loss of generality that the transverse electromagnetic wave propagates along the z direction, so $q_1 = q_2 = 0$ and $\mathbf{q} = (0, 0, q_3)$. Upon performing the Matsubara sum, the transverse component of the electromagnetic response function becomes

$$\begin{aligned} K_{ab}(i\omega_s, \mathbf{q}) &= \frac{e^2}{m|\omega_s|}\left[n\delta_{ab} + \frac{2}{m}\int (d\mathbf{p}) p_a p_b \left(\frac{n(\mathbf{p}-\mathbf{q}/2)}{i\omega_s + \xi(\mathbf{p}-\mathbf{q}/2) - \xi(\mathbf{p}+\mathbf{q}/2)}\right.\right. \\ &\qquad \left.\left. + \frac{n(\mathbf{p}+\mathbf{q}/2)}{-i\omega_s + \xi(\mathbf{p}+\mathbf{q}/2) - \xi(\mathbf{p}-\mathbf{q}/2)}\right)\right] \\ &= \frac{e^2}{m|\omega_s|}\left[n\delta_{ab} + \frac{2}{m}\int (d\mathbf{p}) p_a p_b \left(\frac{n(\mathbf{p})}{i\omega_s + \xi(\mathbf{p}) - \xi(\mathbf{p}+\mathbf{q})}\right.\right. \\ &\qquad \left.\left. + \frac{n(\mathbf{p})}{-i\omega_s + \xi(\mathbf{p}) - \xi(\mathbf{p}+\mathbf{q})}\right)\right], \end{aligned} \tag{10.12}$$

and a, b could be any of the labels 1 and 2. In the numerator, we have a factor $p_a p_b$. Suppose a and b are different, say, 1 and 2 respectively. Then $p_1 p_2 = p^2 \sin^2\theta \cos\phi \sin\phi$, where θ and ϕ are respectively the polar and azimuthal angles of $\mathbf{p}$ reckoned from the z axis along which $\mathbf{q} = (0, 0, q_3)$ lies. Integration over ϕ in Eq. (10.12) gives zero since the rest of the integral depends only on

[1] Upon analytic continuation to real frequencies, the electromagnetic response function is essentially the linear response conductivity at finite frequency and momentum.

θ. So for a nonzero value, we need $a = b$. For $a = b = 1$, the ϕ-dependence is $\cos^2 \phi$, while for $a = b = 2$, it is $\sin^2 \phi$. In both cases, the ϕ integration gives us π.

We first evaluate the momentum dependence of $K_{ab}(i\omega_s, \mathbf{q})$ in the limit $\omega_s = 0$. At zero temperature, the integration is straightforward: one first performs the θ integration followed by the p integration over occupied states, $0 < p < p_F$. The result is

$$K_{ab}(0, \mathbf{q}) = \delta_{ab} \frac{ne^2}{m|\omega_s|} \left[\frac{3}{8}(1 + (|\mathbf{q}|/2p_F)^2) \right. \\ \left. - \frac{3}{16} \frac{(1 - (|\mathbf{q}|/2p_F)^2)^2}{(|\mathbf{q}|/2p_F)} \ln \left| \frac{1 + (|\mathbf{q}|/2p_F)}{1 - (|\mathbf{q}|/2p_F)} \right| \right]. \tag{10.13}$$

Expanding Eq. (10.13) at small $|\mathbf{q}|/2p_F$ one obtains

$$K_{ab}(0, \mathbf{q}) \approx \delta_{ab} \frac{ne^2}{m|\omega_s|} (|\mathbf{q}|/2p_F)^2 = \delta_{ab} \frac{1}{3} \nu(0) \mu_B^2 \frac{\mathbf{q}^2}{|\omega_s|} \equiv \delta_{ab} \frac{1}{3} \chi_{\text{Pauli}} \frac{\mathbf{q}^2}{|\omega_s|}. \tag{10.14}$$

The quantity $\chi_D = \frac{1}{3}\chi_{\text{Pauli}}$ is the Landau diamagnetic susceptibility. This term renormalizes the coefficient of the $\mathbf{q}^2$ term in the action for transverse electromagnetic fluctuations, Eq. (10.8), and the positive sign of $K_{ab}(0, \mathbf{q})$ signifies increased stiffness. Since the coefficient of the $\mathbf{q}^2$ term is just the inverse of the magnetic permittivity μ, the effect of the conduction electrons is to decrease the magnetic permittivity, i.e., a diamagnetic correction. Here it is instructive to compare with the long-wavelength static phase stiffness of a superconductor (given by $|\omega_s|K_{ab}(i\omega_s, \mathbf{q})$ in the $\mathbf{q}$-limit) where it approaches a constant value instead of vanishing like $\mathbf{q}^2$.

Now we present the momentum dependence of the transverse component of $K_{ab}(i\omega_s, \mathbf{q})$ at small but finite frequencies. Performing the integration over the momentum in Eq. (10.12), it readily follows that[2]

$$K_{ab}(i\omega_s, \mathbf{q}) \approx \delta_{ab} \frac{3\pi ne^2}{4p_F} \frac{1}{|\mathbf{q}|} \equiv \delta_{ab} \frac{\pi \nu(0) e^2}{2v_F} \frac{1}{|\mathbf{q}|}. \tag{10.15}$$

Equation (10.15) tells us that the dc conductivity of a clean metal diverges inversely with momentum. For a dirty metal, one would have instead obtained a constant in the limit of zero momentum, $K_{ab}(i\omega_s, \mathbf{q}) \approx (ne^2/m)/(\tau^{-1} + |\omega_s|)$. We are interested precisely in this regime of small momenta and frequencies together with $\omega_s/v_F|\mathbf{q}| < 1$. Using the results of Eq. (10.14) and Eq. (10.15) to simplify the quantity in square brackets in Eq. (10.8) we arrive at the following effective action for the fluctuating transverse electromagnetic fields in a metal:

$$S[\mathbf{A}] \approx \frac{\beta}{2} \sum_{s;a} \int (d\mathbf{q}) A_a(i\omega_s, \mathbf{q}) A_a(-i\omega_s, -\mathbf{q}) \left(\frac{1}{\mu}\mathbf{q}^2 + \frac{|\omega_s|}{\eta|\mathbf{q}|} \right), \tag{10.16}$$

[2]See for example D. Pines and P. Nozières, *The Theory of Quantum Liquids: Volume 1, Normal Fermi Liquids*, Addison Wesley (1988).

with

$$\frac{1}{\mu} = 1 + \frac{1}{3}\chi_{\text{Pauli}} \equiv 1 + \chi_D,$$
$$\text{and } \frac{1}{\eta} = \frac{\pi\nu(0)e^2}{2v_F}. \tag{10.17}$$

For the dirty metal, the factor $1/\eta|\mathbf{q}|$ in Eq. (10.16) should be replaced with the dc conductivity σ. The two-point correlation function corresponding to the action of Eq. (10.16) is

$$D_{ab}(i\omega_s, \mathbf{q}) = \left(\delta_{ab} - \frac{q_a q_b}{\mathbf{q}^2}\right)\frac{1}{\frac{1}{\mu}\mathbf{q}^2 + \frac{|\omega_s|}{\eta|\mathbf{q}|}}, \tag{10.18}$$

and the factor $\delta_{ab} - q_a q_b/\mathbf{q}^2$ enforces $\sum_a q_a D_{ab} = 0$. Equation (10.18) shows that transverse electromagnetic fluctuations in a clean metal get screened over a length scale $\lambda_{\text{skin}} \sim (\eta/\mu|\omega|)^{1/3}$. This is the anomalous skin depth and is different from $\lambda_{\text{skin}} \sim 1/(\mu\sigma|\omega|)^{1/2}$ that we obtained for the dirty metal in Sec. 6.7.

The effective action in Eq. (10.16) is familiar to us as we encountered a similar one earlier (see Sec. 9.6) in the context of Hertz theory, or spin-fluctuation effects in almost ferromagnetic metals. Specifically, Eq. (10.16) corresponds to the classical Gaussian regime, $T > r^{z/2}$, in Hertz theory, with $r = 0$ denoting the quantum critical point. We will use here some of the results obtained in Sec. 9.6.

10.2 Breakdown of the Landau Fermi liquid state

To obtain the thermodynamic potential $\Omega^{(0)}$, we integrate over the transverse electromagnetic fluctuations:

$$\Omega^{(0)} = \frac{V}{\beta}\sum_s \int (d\mathbf{q}) \ln\left[\frac{1}{\mu}\mathbf{q}^2 + \frac{|\omega_s|}{\eta|\mathbf{q}|}\right]. \tag{10.19}$$

We perform the Matsubara sum exactly as it was done following Eq. (9.124) using the contour integral method, which yields

$$\Omega^{(0)}(T) = -2V\int^{\Lambda}(d\mathbf{q})\int_0^{\eta|\mathbf{q}|}\frac{d\omega}{\pi}\coth\left(\frac{\beta\omega}{2}\right)\tan^{-1}\left(\frac{\omega/\eta|\mathbf{q}|}{\mathbf{q}^2/\mu}\right). \tag{10.20}$$

We subtract off the unimportant $T = 0$ contribution:

$$\Omega^{(0)}(T) - \Omega^{(0)}(0) = -2V \int^{\Lambda} (d\mathbf{q}) \int_0^{\eta|\mathbf{q}|} \frac{d\omega}{\pi} \left[\coth\left(\frac{\beta\omega}{2}\right) - 1\right] \times \tan^{-1}\left(\frac{\omega/\eta|\mathbf{q}|}{\mathbf{q}^2/\mu}\right)$$
$$= 4VT \int^{\Lambda} (d\mathbf{q}) \int_0^{\eta|\mathbf{q}|} \frac{d\omega}{\pi} \ln[1 - e^{-\beta\omega}] \times \frac{(\eta|\mathbf{q}|^3/\mu)}{\omega^2 + (\eta|\mathbf{q}|^3/\mu)^2}. \tag{10.21}$$

At low temperatures, the dominant contribution in Eq. (10.21) comes from small frequencies, so we approximate the Lorentzian by $(\mu/\eta|\mathbf{q}|^3$, and replace the ω-cutoff by a smaller value, $\eta|\mathbf{q}|^3/\mu$. The temperature dependence of the thermodynamic potential is then approximately

$$\Omega^{(0)}(T) - \Omega^{(0)}(0) \approx 4VT \int^{\Lambda} (d\mathbf{q}) \int_0^{\eta|\mathbf{q}|^3/\mu} \frac{d\omega}{\pi} \ln[1 - e^{-\beta\omega}] \frac{\mu}{\eta|\mathbf{q}|^3}. \tag{10.22}$$

Then it is easy to see that the singular contribution to the thermodynamic potential at very low temperatures goes as $\Omega^{(0)}(T) - \Omega^{(0)}(0) \approx -VC(\mu/\eta)T^2 \ln(1/T)$, where C is a positive constant.[3] Consequently, the specific heat coefficient diverges,

$$\gamma(T) \sim \frac{\mu}{\eta} \ln(1/T), \tag{10.23}$$

signaling the breakdown of the Landau Fermi liquid state in three dimensions. From Eq. (10.17) above, we have $\eta^{-1} = \frac{\pi\nu(0)e^2}{2v_F} = 3\pi n e^2/(4p_F)$. Introducing back the usual MKS units, the specific heat coefficient has the following singular behavior[4]:

$$\gamma(T) \sim \mu k_B^2 \frac{n}{\epsilon_F} \frac{v_F}{c} \frac{e^2 \mu_0 c}{h} \ln(1/T). \tag{10.24}$$

Equation (10.24) should be compared with the free electron result $\gamma(T) \sim (n/\epsilon_F)k_B^2$. The small dimensionless quantity $(e^2/h)(\mu_0 c) \approx 1.46 \times 10^{-2}$ is the ratio of the impedance of free space ($\approx 377\Omega$) to the quantum of resistance ($\approx 25800\Omega$). The other small dimensionless quantity is the ratio of the Fermi velocity to the speed of propagation of the transverse electromagnetic waves, c. The simultaneous presence of these two small quantities means that one needs to go to incredibly low temperatures before the above contribution of

---

[3]There is also the usual T^2 term that is related to the specific heat of the normal electrons. This is not a singular term even though the (constant) coefficient of T^2 could be large.

[4]M. Yu. Reizer, *Phys. Rev. B* **39**, 1602 (1989).

the transverse electromagnetic fluctuations to the specific heat coefficient becomes comparable to the free electron contribution, $(n/\epsilon_F)k_B^2$, unless of course the magnetic permittivity μ is large. Such a situation arises, for example, in the close proximity of a ferromagnetic transition. So to practically see the breakdown of the Landau Fermi liquid state in a clean metal by transverse electromagnetic fluctuations, it helps if one is close to a ferromagnetic critical point. However it is crucial to note that proximity to a ferromagnetic critical point is not necessary for the existence of the phenomenon, and the Landau Fermi liquid always breaks down at sufficiently low temperatures. Although a similar divergence of the specific heat coefficient is also seen in Hertz theory, there one needs to be in the close proximity of the quantum critical point in order for the spin fluctuations to develop long range correlations. The common element in both cases, of course, is the lack of screening of overdamped particle-hole excitations: magnetization in Hertz theory and current[5] in the present one.

Another possible place where such effects may be searched for are in certain quantum spin-liquid systems where the low energy physics is described by nonrelativistic fermionic spinons coupled to some emergent fluctuating (electromagnetic) gauge field. Since such gauge fields arise from electronic interactions, the speed of light c_{eff} in such models is governed by the scale of electronic interactions, and therefore the quantities v_F/c_{eff}, or even $(e^2/h)(\mu_0 c_{\text{eff}})$ need not be small. It further helps if the systems are, in the language of the effective field theoretical description, close to a "ferromagnetic" critical point.

[5]Hence the name Ampere effects.

Chapter 11

Overscreened Kondo Model and Nozières-Blandin Theory

In our analysis of spin fluctuation effects in nearly ferromagnetic metals in three dimensions, we found that in the "classical" Gaussian regime there is a significant departure from the Landau Fermi liquid behavior, notably in the form of $\ln(T)$ singularities in the leading contribution to the specific heat coefficient $\gamma(T)$. However in the "quantum" regime, Fermi liquid theory continued to be valid. A natural question arises as to whether Fermi liquid theory remains valid even in the presence of weak disorder. For weak homogeneous potential disorder and far from any magnetic instability, the answer is yes – we showed in Ch. 8 that quasiparticles remain well-defined excitations in three dimensional disordered metals. Near the ferromagnetic quantum critical point, Hertz (1976) argued that weak homogeneous potential disorder modifies the dynamical exponent z to four instead of the usual value of three[1], but otherwise the conclusions of spin fluctuation theory should hold.

What if the disorder is in the form of magnetic impurities? This is of course the well-known Kondo problem that we encountered earlier in Ch. 5. The above question was conclusively answered by Wilson who showed, using the numerical renormalization group (NRG) technique he himself developed, that the introduction of a finite but small concentration of spin-$\frac{1}{2}$ magnetic impurities, antiferromagnetically coupled to the metallic host, does not lead to the breakdown of the Fermi liquid state. This is because at low temperatures, the conduction electrons in the metallic host form a many-body (nonmagnetic) singlet state with the impurity spin. Virtual excitations to the triplet state induce a residual repulsive interaction of electrons in the host, which in turn leads to an effective Fermi liquid description. Formation of the many-body

[1] A consequence of spin dynamics becoming diffusive for dirty metals instead of ballistic, as is the case for clean metals.

Kondo singlet is qualitatively different from the usual Lindhard screening of impurities due to polarization of conduction electrons.

The formation of the bound singlet state takes place only if there is a single band of conduction that predominantly interacts with the magnetic impurity. Nozières and Blandin[2] in 1980 considered a hypothetical situation where conduction electrons in more than one band coupled identically with the impurity spin. It turns out that in this case, the attempt by electrons in each of these bands to Kondo "screen" the impurity spin results in "overscreening" with the result that the many-body singlet state is never quite formed. In this chapter, we will study this overscreened Kondo effect using a perturbative scaling technique. Surprisingly, we will also see that the overscreened Kondo effect was discovered nearly a decade before the work of Nozières and Blandin by Larkin and Mel'nikov[3] in their study of the Kondo effect in the vicinity of a ferromagnetic transition.

Before proceeding to analyze the overscreened Kondo model, we will first consider the usual Kondo model as that will bring out the differences between the two, and also give a common framework. In Ch. 5, we were interested in properties of the low temperature Fermi liquid phase. Below we approach the problem from the high temperature end. As we will see, the differences between the usual Kondo physics and overscreened Kondo physics are aapparent even in the high temperature regime.

11.1 Formation of local magnetic moments in a metal

An atom that is magnetic in one metallic host may not retain a magnetic moment when introduced in another metal. For example, an iron atom is not magnetic in a niobium host but develops a local moment when introduced in gold. Anderson[4] in 1961 showed how an otherwise nonmagnetic atom introduced in a metal may develop a local moment. Consider for simplicity the Hamiltonian of a single impurity level below the Fermi level of a metal:

$$H_{\text{atomic}} = H_d + H_U = E_d \sum_{\sigma} n_{d\sigma} + U n_{d\uparrow} n_{d\downarrow}, \tag{11.1}$$

where E_d is the energy of the impurity level, $n_{d\sigma} = d^\dagger_\sigma d_\sigma$ is the number of electrons of spin σ at the impurity site, and U is the Coulomb repulsion when two electrons occupy the impurity site. To form any bound impurity state, we need $E_d < \epsilon_F$. Since for $2E_d + U < 2\epsilon_F$, the level becomes doubly-occupied,

[2]P. Nozières and A. Blandin, *J. Physique* **41**, 193 (1980).

[3]A. I. Larkin and V. I. Mel'nikov, *Sov. Phys. JETP* **34**, 656 (1972) [original version in Russian: *Zh. Eksp. Teor. Fiz.* **61**, 1231 (1971)].

[4]P. W. Anderson, *Phys. Rev.* **124**, 41 (1961).

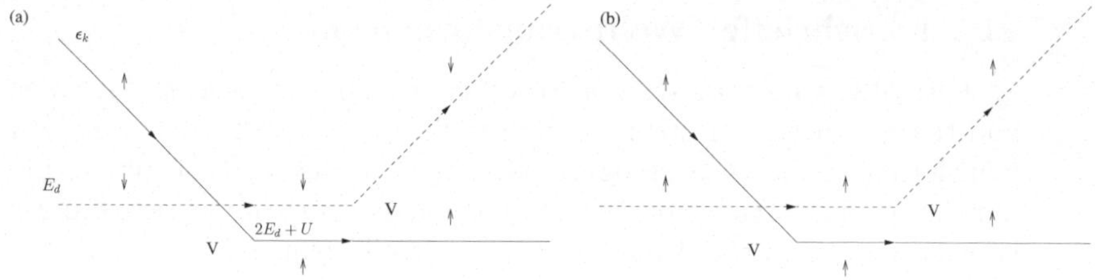

FIGURE 11.1: Origin of antiferromagnetic coupling of local moments and conduction electrons in the Anderson model. In (a), a conduction electron (solid line on the left) with its spin opposite to that of the local moment (dashed line on the left) hops on to the impurity site and hops out with the opposite spin. In (b), the conduction electron has the same spin as that of the local moment and the Pauli principle forbids double occupation of the impurity level by two electrons with the same spin. The second order scattering process in (a) lowers the energy; thus an antiferromagnetic coupling of the local moment and conduction electrons is favored.

we also need $2E_d + U > 2\epsilon_F$. The hybridization of the impurity level with the continuum of energy levels in the Fermi sea is described by the Hamiltonian

$$\begin{aligned} H_A &= H_c + H_{\text{hyb}} + H_{\text{atomic}} \\ &= \sum_{\mathbf{k}\sigma} \epsilon_{\mathbf{k}} c^\dagger_{\mathbf{k}\sigma} c_{\mathbf{k}\sigma} + \mathcal{V} \sum_{\sigma} (c^\dagger_\sigma(0) d_\sigma + \text{h.c.}) + H_{\text{atomic}}. \end{aligned} \tag{11.2}$$

The impurity levels get broadened due to hybridization with the Fermi sea. The broadening can be shown to be $\Delta = \pi \mathcal{V}^2 \nu(0)$ using the Fermi Golden Rule. In order that the singly-occupied level (local moment) is clearly distinguished from the Fermi level as well as the doubly-occupied excited state, we further need $\epsilon_F - E_d \gg \Delta$ and $(U + E_d) \gg \Delta$. The Hamiltonian in Eq. (11.2) is known as the Anderson model.

Next we need to understand how this local moment interacts with the conduction electrons. Fig.11.1 shows two possible second-order processes associated with scattering of a conduction electron from the impurity. The process in (b) is forbidden as it involves double occupation of the impurity site by electrons with the same spin (Pauli principle). The process (a), where the conduction electron and impurity electron have opposite spins, lowers the ground state energy by $\delta E \sim \mathcal{V}^2/(E_d + U - \epsilon_k)$. The energies of the incident and outgoing conduction electrons are close to the Fermi level and this is an example of degenerate perturbation theory.

11.1.1 Schrieffer-Wolff transformation

Formally, we obtain the effective interaction between the local moment and the conduction electrons using the Schrieffer-Wolff transformation[5]. The transformation helps us recognize the origin of the Kondo coupling in virtual ionization of the Anderson impurity atom. To this end we perform a unitary transformation on the Anderson model Hamiltonian H_A,

$$H_K = e^S H_A e^{-S}, \tag{11.3}$$

which eliminates the "off-diagonal" hybridization term in H_A. We choose $S \propto \mathcal{V}$ and denote $H_A = H^{(0)} + H_{\text{hyb}}$. Matching the leading order (in $\mathcal{V}$) terms on both sides of Eq. (11.3) and remembering $H_K = H_K^{(0)} + H_K^{(2)} + \cdots$ does not contain $O(\mathcal{V})$ terms we have

$$\begin{aligned} H_K^{(0)} &= H^{(0)} \\ 0 &= [S, H^{(0)}] + H_{\text{hyb}} \\ H_K^{(2)} &= \frac{1}{2}[S, H_{\text{hyb}}]. \end{aligned} \tag{11.4}$$

The matrix elements of the anti-hermitian matrix S in terms of the unperturbed eigenvalues $E_n^{(0)}$ are $S_{mn} = (H_{\text{hyb}})_{mn}/(E_m^{(0)} - E_n^{(0)})$. The suffixes represent the labels $k\sigma$ of the conduction electrons as well as the impurity spin (if any) and occupation number. Consider first the matrix element corresponding to the creation of a conduction electron $(k\sigma)$ by annihilating an impurity electron with the same spin. The initial impurity state is then either singly or doubly occupied, with corresponding energies E_d and $2E_d + U$ respectively. The matrix element $S_{k\sigma,d\sigma}$ is

$$S_{k\sigma,d\sigma} = \mathcal{V}\left[(1 - n_{d,-\sigma})\frac{1}{\epsilon_k - E_d} + n_{d,-\sigma}\frac{1}{(\epsilon_k + E_d) - (2E_d + U)}\right] c^\dagger_{k\sigma} d_\sigma,$$

from which it follows that

$$S = \mathcal{V}\sum_{k\sigma}\left[\frac{1 - n_{d,-\sigma}}{\epsilon_k - E_d} + \frac{n_{d,-\sigma}}{\epsilon_k - (E_d + U)}\right] c^\dagger_{k\sigma} d_\sigma - \text{h.c.} \tag{11.5}$$

Next we obtain the second order correction $H_K^{(2)}$. The fermionic operators $c_{k\sigma}$ and d_σ satisfy the usual anticommutation relations, $\{c_{k\sigma}, c^\dagger_{k'\sigma'}\} = \delta_{kk'}\delta_{\sigma\sigma'}$, $\{d_\sigma, d^\dagger_{\sigma'}\} = \delta_{\sigma\sigma'}$, and the rest are zero. For $[S_{k\sigma,d\sigma}, H_{\text{hyb}}]$ we need to evaluate four commutators of the form $[c^\dagger_{k\sigma}d_\sigma, c^\dagger_{k'\sigma'}d_{\sigma'}]$, $[d^\dagger_{-\sigma}d_{-\sigma}c^\dagger_{k\sigma}d_\sigma, c^\dagger_{k'\sigma'}d_{\sigma'}]$, $[c^\dagger_{k\sigma}d_\sigma, d^\dagger_{\sigma'}c_{k'\sigma'}]$, $[d^\dagger_{-\sigma}d_{-\sigma}c^\dagger_{k\sigma}d_\sigma, d^\dagger_{\sigma'}c_{k'\sigma'}]$. Using $[AB, C] = A[B, C] + [A, C]B$ and $[ab, cd] = a\{b, c\}d - ac\{b, d\} + \{a, c\}db - c\{a, d\}b$ the above commutators

[5] J. R. Schrieffer and P. A. Wolff, *Phys. Rev.* **149**, 491 (1966).

simplify to

$$
\begin{aligned}
[c^\dagger_{k\sigma}d_\sigma, c^\dagger_{k'\sigma'}d_{\sigma'}] &= 0,\\
[d^\dagger_{-\sigma}d_{-\sigma}c^\dagger_{k\sigma}d_\sigma, c^\dagger_{k'\sigma'}d_{\sigma'}] &= -c^\dagger_{k'\sigma'}d_{-\sigma}c^\dagger_{k\sigma}d_\sigma\delta_{-\sigma,\sigma'},\\
[c^\dagger_{k\sigma}d_\sigma, d^\dagger_{\sigma'}c_{k'\sigma'}] &= c^\dagger_{k\sigma}c_{k'\sigma'}\delta_{\sigma\sigma'} - d^\dagger_{\sigma'}d_\sigma\delta_{\sigma\sigma'}\delta_{kk'},\\
[d^\dagger_{-\sigma}d_{-\sigma}c^\dagger_{k\sigma}d_\sigma, d^\dagger_{\sigma'}c_{k'\sigma'}] &= d^\dagger_{-\sigma}d_{-\sigma}(c^\dagger_{k\sigma}c_{k'\sigma'}\delta_{\sigma\sigma'} - d^\dagger_{\sigma'}d_\sigma\delta_{\sigma\sigma'}\delta_{kk'})\\
&\quad + d^\dagger_{-\sigma}c_{k'\sigma'}c^\dagger_{k\sigma}d_\sigma\delta_{-\sigma,\sigma'}. \qquad (11.6)
\end{aligned}
$$

The remaining ones, $[d^\dagger_\sigma c_{k\sigma}, c^\dagger_{k'\sigma'}d_{\sigma'}]$, $[d^\dagger_{-\sigma}d_{-\sigma}d^\dagger_\sigma c_{k\sigma}, c^\dagger_{k'\sigma'}d_{\sigma'}]$, $[d^\dagger_\sigma c_{k\sigma}, d^\dagger_{\sigma'}c_{k'\sigma'}]$, and $[d^\dagger_{-\sigma}d_{-\sigma}d^\dagger_\sigma c_{k\sigma}, d^\dagger_{\sigma'}c_{k'\sigma'}]$ simplify to

$$
\begin{aligned}
[d^\dagger_\sigma c_{k\sigma}, c^\dagger_{k'\sigma'}d_{\sigma'}] &= d^\dagger_\sigma d_{\sigma'}\delta_{\sigma\sigma'}\delta_{kk'} - c^\dagger_{k'\sigma'}c_{k\sigma}\delta_{\sigma\sigma'},\\
[d^\dagger_{-\sigma}d_{-\sigma}d^\dagger_\sigma c_{k\sigma}, c^\dagger_{k'\sigma'}d_{\sigma'}] &= d^\dagger_{-\sigma}d_{-\sigma}(d^\dagger_\sigma d_{\sigma'}\delta_{\sigma\sigma'}\delta_{kk'} - c^\dagger_{k'\sigma'}c_{k\sigma}\delta_{\sigma\sigma'})\\
&\quad - c^\dagger_{k'\sigma'}d_{-\sigma}d^\dagger_\sigma c_{k\sigma}\delta_{-\sigma,\sigma'},\\
[d^\dagger_\sigma c_{k\sigma}, d^\dagger_{\sigma'}c_{k'\sigma'}] &= 0,\\
[d^\dagger_{-\sigma}d_{-\sigma}d^\dagger_\sigma c_{k\sigma}, d^\dagger_{\sigma'}c_{k'\sigma'}] &= d^\dagger_{-\sigma}c_{k'\sigma'}d^\dagger_\sigma c_{k\sigma}\delta_{-\sigma,\sigma'}. \qquad (11.7)
\end{aligned}
$$

The resulting expression for $H_K^{(2)}$ is simplified by introducing the parameters

$$
\begin{aligned}
W(k,k') &= \frac{\mathcal{V}^2/2}{\epsilon_k - E_d} + k \leftrightarrow k',\\
J(k,k') &= -\frac{\mathcal{V}^2 U}{2(\epsilon_k - E_d)(\epsilon_k - (E_d + U))} + k \leftrightarrow k', \qquad (11.8)
\end{aligned}
$$

that leads to

$$
H_K^{(2)} = H_{\text{ex}} + \delta H_c + \delta H_{\text{atomic}} + H_{\text{sc}}, \qquad (11.9)
$$

where

$$
\begin{aligned}
H_{\text{ex}} &= \sum_{kk',\alpha\beta,\gamma\delta} J(k,k')\left(c^\dagger_{k\alpha}\sigma_{\alpha\beta}c_{k'\beta}\right)\left(d^\dagger_\gamma \frac{\boldsymbol{\sigma}_{\gamma\delta}}{2} d_\delta\right),\\
\delta H_c &= \sum_{kk',\sigma}\left(W(k,k') - \frac{1}{2}J(k,k')n_d\right)c^\dagger_{k\sigma}c_{k'\sigma},\\
\delta H_{\text{atomic}} &= -\sum_{k\sigma}(W(k,k') - J(k,k')n_{d,-\sigma})\,n_{d\sigma},\\
\delta H_{\text{sc}} &= \frac{1}{2}\sum_{kk',\sigma} J(k,k')(c^\dagger_{k',-\sigma}c^\dagger_{k\sigma}d_\sigma d_{-\sigma} + \text{h.c.}). \qquad (11.10)
\end{aligned}
$$

The conduction electrons that hybridize with the impurity come from the vicinity of the Fermi energy. We may thus replace the parameters W and

J in Eq. (11.8) by their values at $k = k' = k_F$. Since $\epsilon_F - E_d > 0$, $W = \mathcal{V}^2/(\epsilon_F - E_d) > 0$. From our discussion above on local moment formation, $2E_d + U > 2\epsilon_F$, or $E_d - \epsilon_F > \epsilon_F - (E_d + U)$. Since $E_d - \epsilon_F < 0$, we thus have $J = -\mathcal{V}^2 U/[(\epsilon_F - E_d)(\epsilon_F - E_d - U] > 0$. Let us now examine the terms in Eq. (11.10). $\delta H_{\rm sc}$ changes the impurity occupation by two and is relevant when impurity occupation $n_d = 2$ or 0. We are in the local moment regime which corresponds to the $n_d = 1$ sector of the unperturbed Hamiltonian $H^{(0)}$. Thus $\delta H_{\rm sc}$ may be ignored in the rest of the discussion. $\delta H_{\rm atomic}$ redefines E_d and U and may be absorbed in an effective $H_{\rm atomic}$. For $n_d = 1$, δH_c describes potential (spin-independent) scattering off the impurity and leads to a small renormalization of the quasiparticle energies ϵ_k. $\delta H_{\rm ex}$ describes the *antiferromagnetic* coupling of the impurity moment and the local conduction electron spin density. The properties of the local moment regime are effectively described by

$$H_K = H_c + H_{\rm ex} \equiv \sum_{\mathbf{k}\sigma} \epsilon_{\mathbf{k}} c^\dagger_{\mathbf{k}\sigma} c_{\mathbf{k}\sigma} + \frac{1}{V} \sum_{\mathbf{k}\mathbf{k}',\alpha\beta} J(\mathbf{k},\mathbf{k}') c^\dagger_{\mathbf{k}\alpha} \boldsymbol{\sigma}_{\alpha\beta} c_{\mathbf{k}'\beta} \cdot \mathbf{S}, \quad (11.11)$$

where $\mathbf{S}$ is the local impurity spin and V is the volume of the system. Note that now $J(\mathbf{k},\mathbf{k}')$has the dimensions of energy times volume. Eq. (11.11) is the Kondo or $s-d$ model that we introduced in Ch. 5.

Kondo[6] in 1964 provided an explanation for the resistance upturn in metals with magnetic impurities by showing that spin-flip scattering of conduction electrons from the magnetic impurity leads to $\ln(1/T)$ corrections to the resistivity. We present a sketch of Kondo's explanation below.

11.2 Perturbation theory for single-channel Kondo model

Drude's formula $\rho = m/ne^2\tau_{\rm tr}$ relates the resistivity ρ of a nearly free electron conductor with the momentum relaxation rate (or transport scattering rate), $1/\tau_{\rm tr}$, of the charge carriers. If there are a number of independent mechanisms for relaxation of momentum, $1/\tau_{\rm tr}$ is the sum of the momentum relaxation rates from the individual mechanisms (Matthiesen rule). We are interested in the contribution to resistivity from scattering by the impurity local moments. The momentum relaxation rate $\tau_{\rm tr}^{-1}(\mathbf{k})$ for an electron with wavevector $\mathbf{k}$ is related to the impurity scattering rate $\tau_{\rm imp}^{-1}(\mathbf{k},\mathbf{k}')$ through

$$\tau_{\rm tr}^{-1}(\mathbf{k}) = N_{\rm imp} \sum_{\mathbf{k}'} (1 - \cos\theta_{\mathbf{k}\mathbf{k}'}) \tau_{\rm imp}^{-1}(\mathbf{k},\mathbf{k}'), \quad (11.12)$$

[6] J. Kondo, *Prog. Theor. Phys.* **32**, 37 (1964).

where $N_{\rm imp}$ is the number of impurities. From the Fermi Golden Rule,

$$\tau_{\rm imp}^{-1}(\mathbf{k}, \mathbf{k}') = \frac{2\pi}{\hbar} {\rm tr}|T_{\mathbf{k}\mathbf{k}'}|^2 \delta(\epsilon_{\mathbf{k}'} - \epsilon_{\mathbf{k}}), \tag{11.13}$$

where T is the $T-$matrix defined by

$$T|\psi^{(0)}\rangle = H_{\rm ex}|\psi\rangle, \tag{11.14}$$

and the trace is over spin indices. $|\psi^{(0)}\rangle$ and $|\psi\rangle$ refer to eigenstates of the free electron Hamiltonian H_c and the Kondo Hamiltonian $H_K = H_c + H_{\rm ex}$ respectively. The matrix elements of T can be obtained as follows (see Ch. 4 for details). First note that the formal solution of $H_K|\psi\rangle = E|\psi\rangle$ is given by the Lippmann-Schwinger formula, $|\psi\rangle = |\psi^{(0)}\rangle + (E - H_c + i\eta)^{-1} H_{\rm ex}|\psi\rangle$. Then from Eq. (11.14) it follows that $T = H_{\rm ex} + H_{\rm ex}(E - H_c + i\eta)^{-1}T$. This leads us to the following expansion for the $T-$matrix:

$$\begin{aligned} T = H_{\rm ex} &+ H_{\rm ex}\frac{1}{E - H_c + i\eta}H_{\rm ex} \\ &+ H_{\rm ex}\frac{1}{E - H_c + i\eta}H_{\rm ex}\frac{1}{E - H_c + i\eta}H_{\rm ex} + \cdots . \end{aligned} \tag{11.15}$$

The ground state of the unperturbed system is the Fermi sea filled up to the chemical potential. We take this as our vacuum state. To calculate the matrix elements of T, we may use Wick's theorem after noting that the action of fermion creation and annihilation operators is slightly different from the usual case. For our ground state $|\Omega\rangle = \prod_{\sigma, k\leq k_F} c^\dagger_{\mathbf{k}\sigma}|0\rangle$ (Fermi sea filled up to the chemical potential), it is easy to see that for hole creation operators, $c^\dagger_{\mathbf{k}\sigma}|\Omega\rangle = 0$, $k < k_F$ and for particle annihilation operators, $c_{\mathbf{k}\sigma}|\Omega\rangle = 0$, $k > k_F$. Normal ordering is also correspondingly redefined: all the particle annihilation and hole creation operators are to be moved to the right and all the particle creation and hole annihilation operators are to be moved to the left. Contraction, or averaging, is over $|\Omega\rangle$ instead of $|0\rangle$. Thus $\langle\Omega|c^\dagger_{\mathbf{k}\sigma}c_{\mathbf{k}'\sigma'}|\Omega\rangle = \delta_{\mathbf{k}\mathbf{k}'}\delta_{\sigma\sigma'}n_{\mathbf{k}} \equiv \delta_{\mathbf{k}\mathbf{k}'}\delta_{\sigma\sigma'}\theta(k_F - k)$ and $H_c|\Omega\rangle = \sum_{k\leq k_F,\sigma}\epsilon_{\mathbf{k}}|\Omega\rangle \equiv E_0|\Omega\rangle$ (E_0 is the ground state energy of the entire system of electrons). We can also verify the following two relations straightforwardly:

$$\begin{aligned} H_c c^\dagger_{\mathbf{k}\sigma} &= c^\dagger_{\mathbf{k}\sigma}(\epsilon_{\mathbf{k}} + H_c), \\ H_c c_{\mathbf{k}\sigma} &= c_{\mathbf{k}\sigma}(-\epsilon_{\mathbf{k}} + H_c). \end{aligned} \tag{11.16}$$

From Eq. (11.16), it follows that

$$\begin{aligned} \frac{1}{E - H_c}c^\dagger_{\mathbf{k}\sigma} &= c^\dagger_{\mathbf{k}\sigma}\frac{1}{E - (\epsilon_{\mathbf{k}} + H_c)}, \\ \frac{1}{E - H_c}c_{\mathbf{k}\sigma} &= c_{\mathbf{k}\sigma}\frac{1}{E - (-\epsilon_{\mathbf{k}} + H_c)}. \end{aligned} \tag{11.17}$$

Now we are in a position to evaluate the matrix elements of T order by order in H_{ex} as shown in Eq. (11.15). We need to calculate $\langle \mathbf{k}\sigma|T|\mathbf{k}'\sigma'\rangle$. We make a simplifying assumption that $J(\mathbf{k},\mathbf{k}') = J$ in a band of width $2D$ centered at the Fermi energy and zero outside. The leading order contribution to the T−matrix is

$$T^{(1)}_{\mathbf{k}\sigma,\mathbf{k}'\sigma'} = (J/V)\sum_{\mathbf{q},\mathbf{q}',\alpha\beta} \mathbf{S}\cdot\boldsymbol{\sigma}_{\alpha\beta}\langle\Omega|c_{\mathbf{k}\sigma}c^{\dagger}_{\mathbf{q}\alpha}c_{\mathbf{q}'\beta}c^{\dagger}_{\mathbf{k}'\sigma'}|\Omega\rangle.$$

Using Wick's theorem to perform the averaging we obtain two contributions of the form $\langle c_{\mathbf{k}\sigma}c^{\dagger}_{\mathbf{q}\alpha}\rangle\langle c_{\mathbf{q}'\beta}c^{\dagger}_{\mathbf{k}'\sigma'}\rangle$ and $\langle c_{\mathbf{k}\sigma}c^{\dagger}_{\mathbf{k}'\sigma'}\rangle\langle c^{\dagger}_{\mathbf{q}\alpha}c_{\mathbf{q}'\beta}\rangle$. The second term is a disconnected diagram and we discard it since it can be absorbed in a redefinition of the chemical potential. The first term gives us

$$T^{(1)}_{\mathbf{k}\sigma,\mathbf{k}'\sigma'} = (J/V)\mathbf{S}\cdot\boldsymbol{\sigma}_{\sigma\sigma'}(1-n_{\mathbf{k}})(1-n_{\mathbf{k}'}). \tag{11.18}$$

Consider now the second order contribution to the T−matrix:

$$\begin{aligned}
T^{(2)}_{\mathbf{k}\sigma,\mathbf{k}'\sigma'} &= (J/V)^2 \sum_{\mathbf{q}\mathbf{q}',\mathbf{p}\mathbf{p}',\alpha\beta\gamma\delta} (\mathbf{S}\cdot\boldsymbol{\sigma}_{\alpha\beta})(\mathbf{S}\cdot\boldsymbol{\sigma}_{\gamma\delta}) \\
&\quad \times \langle\Omega|c_{\mathbf{k}\sigma}c^{\dagger}_{\mathbf{q}\alpha}c_{\mathbf{q}'\beta}\frac{1}{E-H_c}c^{\dagger}_{\mathbf{p}\gamma}c_{\mathbf{p}'\delta}c^{\dagger}_{\mathbf{k}'\sigma'}|\Omega\rangle \\
&= (J/V)^2 \sum_{\mathbf{q}\mathbf{q}',\mathbf{p}\mathbf{p}',\alpha\beta\gamma\delta} (\mathbf{S}\cdot\boldsymbol{\sigma}_{\alpha\beta})(\mathbf{S}\cdot\boldsymbol{\sigma}_{\gamma\delta})\frac{1}{(E-E_0)-(\epsilon_{\mathbf{p}}-\epsilon_{\mathbf{p}'}+\epsilon_{\mathbf{k}'})} \\
&\quad \times \langle\Omega|c_{\mathbf{k}\sigma}c^{\dagger}_{\mathbf{q}\alpha}c_{\mathbf{q}'\beta}c^{\dagger}_{\mathbf{p}\gamma}c_{\mathbf{p}'\delta}c^{\dagger}_{\mathbf{k}'\sigma'}|\Omega\rangle.
\end{aligned} \tag{11.19}$$

Performing Wick's averaging and discarding disconnected terms as before, we are left with four contributions. Two of them are

$$\langle c_{\mathbf{k}\sigma}c^{\dagger}_{\mathbf{p}\gamma}\rangle\langle c^{\dagger}_{\mathbf{q}\alpha}c_{\mathbf{q}'\beta}\rangle\langle c_{\mathbf{p}'\delta}c^{\dagger}_{\mathbf{k}'\sigma'}\rangle = \delta_{\mathbf{p}\mathbf{k}}\delta_{\sigma\gamma}\delta_{\mathbf{q}\mathbf{q}'}\delta_{\alpha\beta}\delta_{\mathbf{p}'\mathbf{k}'}\delta_{\delta\sigma'}n_{\mathbf{q}}(1-n_{\mathbf{k}})(1-n_{\mathbf{k}'})$$

and

$$\langle c_{\mathbf{k}\sigma}c^{\dagger}_{\mathbf{q}\alpha}\rangle\langle c^{\dagger}_{\mathbf{p}\gamma}c_{\mathbf{p}'\delta}\rangle\langle c_{\mathbf{q}'\beta}c^{\dagger}_{\mathbf{k}'\sigma'}\rangle = \delta_{\mathbf{q}\mathbf{k}}\delta_{\sigma\alpha}\delta_{\mathbf{p}\mathbf{p}'}\delta_{\gamma\delta}\delta_{\mathbf{q}'\mathbf{k}'}\delta_{\beta\sigma'}n_{\mathbf{p}}(1-n_{\mathbf{k}})(1-n_{\mathbf{k}'}).$$

They make no contribution to $T^{(2)}$ since they involve the traces of Pauli matrices of the form $\sum_{\gamma}(\mathbf{S}\cdot\boldsymbol{\sigma}_{\gamma\gamma})$ which vanish identically. That leaves us with the following two second order contributions:

$$\begin{aligned}
&\langle c_{\mathbf{k}\sigma}c^{\dagger}_{\mathbf{q}\alpha}\rangle\langle c_{\mathbf{q}'\beta}c^{\dagger}_{\mathbf{p}\gamma}\rangle\langle c_{\mathbf{p}'\delta}c^{\dagger}_{\mathbf{k}'\sigma'}\rangle \\
&\qquad = \delta_{\mathbf{q}\mathbf{k}}\delta_{\sigma\alpha}\delta_{\mathbf{p}\mathbf{q}'}\delta_{\beta\gamma}\delta_{\mathbf{p}'\mathbf{k}'}\delta_{\delta\sigma'}(1-n_{\mathbf{p}})(1-n_{\mathbf{k}})(1-n_{\mathbf{k}'})
\end{aligned}$$

and

$$\begin{aligned}
&-\langle c_{\mathbf{k}\sigma}c^{\dagger}_{\mathbf{p}\gamma}\rangle\langle c^{\dagger}_{\mathbf{q}\alpha}c_{\mathbf{p}'\delta}\rangle\langle c_{\mathbf{q}'\beta}c^{\dagger}_{\mathbf{k}'\sigma'}\rangle \\
&\qquad = -\delta_{\mathbf{p}\mathbf{k}}\delta_{\sigma\gamma}\delta_{\mathbf{q}\mathbf{p}'}\delta_{\alpha\delta}\delta_{\mathbf{k}'\mathbf{q}'}\delta_{\beta\sigma'}n_{\mathbf{p}'}(1-n_{\mathbf{k}})(1-n_{\mathbf{k}'}).
\end{aligned}$$

Note the negative sign in the above term – this arises from the fact that $c_{\mathbf{k}\sigma}c^\dagger_{\mathbf{p}\gamma}c^\dagger_{\mathbf{q}\alpha}c_{\mathbf{p}'\delta}c_{\mathbf{q}'\beta}c^\dagger_{\mathbf{k}'\sigma'}$ involves an odd permutation of the original sequence of fermions in Eq. (11.19). Thus we have

$$\begin{aligned} T^{(2)}_{\mathbf{k}\sigma,\mathbf{k}'\sigma'} &= (J/V)^2\sum_{\mathbf{p}\beta}(\mathbf{S}\cdot\boldsymbol{\sigma}_{\sigma\beta})(\mathbf{S}\cdot\boldsymbol{\sigma}_{\beta\sigma'})\frac{(1-n_{\mathbf{p}})(1-n_{\mathbf{k}})(1-n_{\mathbf{k}'})}{(E-E_0)-\epsilon_{\mathbf{p}}} \\ &\quad -(J/V)^2\sum_{\mathbf{p}'\alpha}(\mathbf{S}\cdot\boldsymbol{\sigma}_{\alpha\sigma'})(\mathbf{S}\cdot\boldsymbol{\sigma}_{\sigma\alpha})\frac{n_{\mathbf{p}'}(1-n_{\mathbf{k}})(1-n_{\mathbf{k}'})}{(E-E_0)-(\epsilon_{\mathbf{k}}-\epsilon_{\mathbf{p}'}+\epsilon_{\mathbf{k}'})}. \end{aligned}$$

We perform the momentum summations using the approximation $2\sum_{\mathbf{p}} = 2V\int(d\mathbf{p}) \approx \nu(0)V\int d\epsilon_{\mathbf{p}}$ near the Fermi surface and recalling the Kondo coupling J is constant in the energy band $[-D+\epsilon_F, D+\epsilon_F]$ and zero outside the band:

$$\begin{aligned} T^{(2)}_{\mathbf{k}\sigma,\mathbf{k}'\sigma'} &= -\frac{J^2\nu(0)}{2V}\Bigg(\sum_{\beta}(\mathbf{S}\cdot\boldsymbol{\sigma}_{\sigma\beta})(\mathbf{S}\cdot\boldsymbol{\sigma}_{\beta\sigma'}) \\ &\qquad \times(1-n_{\mathbf{k}})(1-n_{\mathbf{k}'})\ln\left|\frac{D-(E-E_0-\epsilon_F)}{(E-E_0)-\epsilon_F}\right|\Bigg) \\ &\quad -\frac{J^2\nu(0)}{2V}\Bigg(\sum_{\alpha}(\mathbf{S}\cdot\boldsymbol{\sigma}_{\alpha\sigma'})(\mathbf{S}\cdot\boldsymbol{\sigma}_{\sigma\alpha}) \\ &\qquad \times(1-n_{\mathbf{k}})(1-n_{\mathbf{k}'})\ln\left|\frac{(E-E_0)-\epsilon_F}{(E-E_0-\epsilon_F)-D}\right|\Bigg). \end{aligned}$$

We have replaced $\epsilon_{\mathbf{k}}$ and $\epsilon_{\mathbf{k}'}$ by ϵ_F since low energy scattering takes place close to the Fermi surface. For the same reason, $E-E_0 \approx \epsilon_F$. If the bandwidth $2D$ over which J is significant is large enough such that $D \gg (E-E_0-\epsilon_F)$, the second order contribution to the $T-$matrix further simplifies to

$$\begin{aligned} T^{(2)}_{\mathbf{k}\sigma,\mathbf{k}'\sigma'} &\approx -\frac{J^2\nu(0)}{2V}(1-n_{\mathbf{k}})(1-n_{\mathbf{k}'})\ln\left|\frac{D}{(E-E_0)-\epsilon_F}\right|\sum_{ij}S_iS_j[\sigma^i,\sigma^j]_{\sigma\sigma'} \\ &= -i\frac{J^2\nu(0)}{2V}(1-n_{\mathbf{k}})(1-n_{\mathbf{k}'})\ln\left|\frac{D}{(E-E_0)-\epsilon_F}\right|\sum_{ij}S_iS_j\sigma^k_{\sigma\sigma'}\epsilon_{ijk} \\ &= \frac{J^2\nu(0)}{2V}(1-n_{\mathbf{k}})(1-n_{\mathbf{k}'})\ln\left|\frac{D}{(E-E_0)-\epsilon_F}\right|(\mathbf{S}\cdot\boldsymbol{\sigma})_{\sigma\sigma'}. \end{aligned} \quad (11.20)$$

We made use of the identity $[S_i, S_j] = i\sum_l \epsilon_{ijl}S_l$. Had the impurity spin been classical, or the scattering been diagonal in spin, this commutator would have vanished, and so would $T^{(2)}$. In other words, the finiteness of $T^{(2)}$ is a purely quantum mechanical phenomenon and is associated with spin-flip scattering. Note that $T^{(2)}$ diverges logarithmically as the excitation energy $(E-E_0)$ approaches the Fermi energy ϵ_F. The logarithmic divergence represents contributions from virtual excitations to intermediate states that may be far from the Fermi energy. Clearly, states far from the Fermi energy make a significant contribution to the $T-$matrix. Comparing $T^{(2)}$ with the first order contribution $T^{(1)}$ in Eq. (11.18), we see that even when $J\nu(0) \ll 1$, the second order contribution can become comparable to the first order contribution $T^{(1)}$ for sufficiently small excitation energies. Perturbation theory breaks down for $(E - E_0) = T_K \sim D\exp[-1/J\nu(0)]$. The exponentially small energy scale T_K is known in the literature as the Kondo temperature. At finite temperatures, the logarithmic divergence at the Fermi energy is cut off and has a $\ln|D/T|$ behavior. For such finite temperature scattering near the Fermi surface, we can approximate the expression for τ_{tr}^{-1} in Eq. (11.13) by

$$\begin{aligned}\tau_{\text{tr}}^{-1} &\approx c_{\text{imp}}\frac{2\pi(J + J^2\nu(0)\ln|D/T|)^2}{\hbar}\text{tr}(\mathbf{S}\cdot\boldsymbol{\sigma})^2 \\ &\quad\times\int\frac{2\pi k^2 dk}{(2\pi)^3}\sin\theta d\theta\,\delta(\epsilon_F - \epsilon_{\mathbf{k}})(1-\cos\theta) \\ &\approx c_{\text{imp}}\frac{3\pi\nu(0)J^2}{2\hbar}\left(1 + 2J\nu(0)\ln\left|\frac{D}{T}\right|\right), \end{aligned} \tag{11.21}$$

where c_{imp}/V is the impurity concentration we used $\text{tr}(\mathbf{S}\cdot\boldsymbol{\sigma})^2 = 2S(S+1) = 3/2$ and $2\int(d^3k')\delta(\epsilon_F - \epsilon_{k'}) = \nu(0)$. Consequently, the contribution to the resistivity, $\rho_{\text{imp}} = (m/ne^2)\tau_{\text{tr}}^{-1}$, has a $\ln(1/T)$ upturn.

Such logarithmic corrections show up in a number of other properties such as the impurity susceptibility and specific heat. For example, consider the magnetization $\langle S_z\rangle$ in the presence of a small external magnetic field h ($\ll T$) along the $z-$direction. The noninteracting part of the Hamiltonian now has the additional Zeeman terms $-\mu_B h\sum_{\mathbf{k}}(c^\dagger_{\mathbf{k}\uparrow}c_{\mathbf{k}\uparrow} - c^\dagger_{\mathbf{k}\downarrow}c_{\mathbf{k}\downarrow}) - g\mu_B hS_z$, where $g = 2$ is the electron gyromagnetic ratio. For the isolated impurity, $\langle S_z^{(0)}\rangle = (1/Z)\sum_{m=\pm1/2} me^{2\mu_B h\beta m}$. The Kondo coupling H_{ex} is equivalent to an effective magnetic field $\mathbf{h}_{\text{ex}} = -(J/Vg\mu_B)\sum_{\mathbf{k}\mathbf{k}',\alpha\beta}\langle c^\dagger_{\mathbf{k}\alpha}\boldsymbol{\sigma}_{\alpha\beta}c_{\mathbf{k}'\beta}\rangle$ on the impurity spin. The change in impurity energy when a small field h is turned on is

$$\delta E_h = -(g\mu_B/2)\langle(hS_z + \mathbf{h}_{\text{ex}}\cdot\mathbf{S}\rangle.$$

The evaluation of $\langle\mathbf{h}_{\text{ex}}\cdot\mathbf{S}\rangle$ proceeds much like the $T-$matrix; however, one needs to take care of the different occupation probabilities for the two spin species ($c^\dagger_{\mathbf{k}\sigma}c_{\mathbf{k}\sigma}|\Omega\rangle = n_{\mathbf{k}\sigma}|\Omega\rangle$). One can show that the impurity susceptibility $\chi_{\text{imp}} = g\mu_B c_{\text{imp}}\langle S_z\rangle/h$ is

$$\chi_{\text{imp}} \equiv c_{\text{imp}}(g_{\text{eff}}\mu_B)^2/4T \approx c_{\text{imp}}(\mu_B^2/T)\left[1 - J\nu(0) - (J\nu(0))^2\ln\left|\frac{D}{T}\right|\right]. \tag{11.22}$$

Note that as $T \to T_K = D\exp[-1/J\nu(0)]$, the effective gyromagnetic ratio of the impurity spin, g_{eff} tends to vanish. Physically, the conduction electrons screen the impurity moment which tends to vanish as the temperature approaches T_K.

Instead of calculating the corrections to different physical quantities over and over again, we can reduce our efforts considerably by noting that there is essentially one parameter in the problem, namely, the ratio of the Kondo coupling to the conduction electron bandwidth. Following the perturbative RG philosophy discussed earlier in the context of Hertz theory, it makes sense to study the scaling of this parameter and express the physical quantities in terms of this effective parameter. We will use a simplified RG scheme, namely, Anderson's "poor man's scaling". The difference from the RG scheme we employed in our treatment of Hertz theory is that poor man's scaling dispenses with the restoring of the momentum cutoff step, and simply tracks the flow of the Kondo parameter in response to elimination of fast degrees of freedom. The rationale for not keeping track of the renormalizations of the conduction electron properties is that a single (or a small concentration) of magnetic impurities will not affect the properties of the metal host.

11.3 Poor man's scaling

At low energies, one would like to study effective Hamiltonians that involve a small number of states close to the ground state. Nevertheless, the parameters in the effective Hamiltonian might be dependent on much larger energy scales. The idea behind Anderson's poor man's scaling[7] is very simple – we want to obtain an effective low energy model defined in a smaller energy band around the Fermi surface which has the same low energy physics as the original Hamiltonian which was defined in a larger energy band.

We begin with the Lippmann-Schwinger equation for the $T-$matrix,

$$T = H_{\rm ex} + H_{\rm ex}\frac{1}{E - H_c + i\eta}T. \tag{11.23}$$

We now separate the excitations of the scattered electrons into those whose energies lie within a band $[-D'+\epsilon_F, D'+\epsilon_F] = [-D+|\delta D|+\epsilon_F, D-|\delta D|+\epsilon_F]$ and those with energies in the outlying bands $[-D+\epsilon_F, -D'+\epsilon_F]$, $[D'+\epsilon_F, D+\epsilon_F]$. This we write formally as

$$T = H_{\rm ex} + H_{\rm ex}(1 - P_{\delta D})R_0 T + H_{\rm ex}P_{\delta D}R_0 T, \tag{11.24}$$

where $P_{\delta D}$ projects onto *states with one or more particles in the outlying bands* and R_0 is the resolvent $1/(E - H_c + i\eta)$. Upon using Eq. (11.23) on the right

[7]P. W. Anderson, *J. Phys. C: Solid St. Phys* **3**, 2436 (1970).

hand side of Eq. (11.24),

$$\begin{aligned}
T &= H_{\rm ex} + H_{\rm ex}(1 - P_{\delta D})R_0 T \\
&\quad + H_{\rm ex}P_{\delta D}R_0\left(H_{\rm ex} + H_{\rm ex}[(1 - P_{\delta D})R_0 + P_{\delta D}R_0]T\right) \\
&= (H_{\rm ex} + H_{\rm ex}P_{\delta D}R_0 H_{\rm ex}) + (H_{\rm ex} + H_{\rm ex}P_{\delta D}R_0 H_{\rm ex})(1 - P_{\delta D})R_0 T \\
&\quad + H_{\rm ex}P_{\delta D}R_0 H_{\rm ex}P_{\delta D}R_0 T \\
&= H'_{\rm ex} + H'_{\rm ex}(1 - P_{\delta D})T + H_{\rm ex}P_{\delta D}R_0 H_{\rm ex}P_{\delta D}R_0 T.
\end{aligned} \tag{11.25}$$

The third term is of the order of $H_{\rm ex}^2 \nu(0)^2 \delta D^2/D^2$ (higher order in δD) and we drop it. Eq. 11.25 says that the $T-$matrix of the original Hamiltonian (defined in $[-D + \epsilon_F, D + \epsilon_F]$) is the same as the $T-$matrix of a new Hamiltonian defined in the reduced band $[-D' + \epsilon_F, D' + \epsilon_F]$ with a renormalized Kondo interaction

$$H'_{\rm ex} = H_{\rm ex} + H_{\rm ex}P_{\delta D}\frac{1}{E - H_c + i\eta}H_{\rm ex}. \tag{11.26}$$

Eq. (11.26) is similar to the second order perturbation expansion of the $T-$matrix that we have studied earlier.

We shall study a more general anisotropic Kondo interaction,

$$\begin{aligned}
H_{\rm ex} = (J_z/V)\sum_{\mathbf{k}\mathbf{k}',\alpha\beta} c^\dagger_{\mathbf{k}\alpha}\sigma^z_{\alpha\beta}c_{\mathbf{k}'\beta}S_z + (J_\perp/V)\sum_{\mathbf{k}\mathbf{k}'} c^\dagger_{\mathbf{k}\uparrow}c_{\mathbf{k}'\downarrow}S^- \\
+ (J_\perp/V)\sum_{\mathbf{k}\mathbf{k}'} c^\dagger_{\mathbf{k}\downarrow}c_{\mathbf{k}'\uparrow}S^+,
\end{aligned} \tag{11.27}$$

where the transverse coupling $J_\perp$ may be different from the longitudinal coupling J_z. Consider the second order term in Eq. (11.26):

$$\begin{aligned}
H'^{(2)}_{\rm ex} = \frac{1}{V^2}\sum_{\mathbf{k}\mathbf{k}',\mathbf{q}\mathbf{q}',\alpha\beta\gamma\delta,ij} J_i J_j \Bigg[S_i S_j \sigma^i_{\alpha\beta}\sigma^j_{\gamma\delta}c^\dagger_{\mathbf{k}\alpha}c_{\mathbf{q}\beta}\frac{1}{E - H_c}c^\dagger_{\mathbf{q}'\gamma}c_{\mathbf{k}'\delta} \\
+ S_j S_i \sigma^j_{\gamma\delta}\sigma^i_{\alpha\beta}c^\dagger_{\mathbf{q}'\gamma}c_{\mathbf{k}'\delta}\frac{1}{E - H_c}c^\dagger_{\mathbf{k}\alpha}c_{\mathbf{q}\beta}\Bigg],
\end{aligned} \tag{11.28}$$

where $\mathbf{k}$, $\mathbf{k}'$ belong to the reduced band $[-D' + \epsilon_F, D' + \epsilon_F]$ and $\mathbf{q}$, $\mathbf{q}'$ belong to the outlying bands described above. For simplicity of notation and without loss of generality, for the rest of the chapter we set the ground state energy $E_0 = 0$ and all electron energies are measured with respect to the Fermi energy. Thus the $\epsilon_{\mathbf{k}}$ now lie in the range $[-D', D']$ and the outlying bands are $[-D, -D + |\delta D|]$ and $[D - |\delta D|, D]$. In Eq. (11.28), we move the fermion operators on the right through the resolvent and obtain

$$\begin{aligned}
H'^{(2)}_{\rm ex} = \frac{1}{V^2}\sum_{\mathbf{k}\mathbf{k}',\mathbf{q},\alpha\beta\delta,ij} J_i J_j \sigma^i_{\alpha\beta}\sigma^j_{\beta\delta}\Bigg[S_i S_j c^\dagger_{\mathbf{k}\alpha}c_{\mathbf{k}'\delta}\frac{(1 - n_{\mathbf{q}})}{E - (\epsilon_{\mathbf{q}} - \epsilon_{\mathbf{k}'})} \\
+ S_j S_i c_{\mathbf{k}'\delta}c^\dagger_{\mathbf{k}\alpha}\frac{n_{\mathbf{q}}}{E - (\epsilon_{\mathbf{k}} - \epsilon_{\mathbf{q}})}\Bigg].
\end{aligned} \tag{11.29}$$

Here we have replaced $c_{\mathbf{q}\beta}c^{\dagger}_{\mathbf{q}'\gamma} = \delta_{\mathbf{q}\mathbf{q}'}\delta_{\beta\gamma}(1-n_{\mathbf{q}})$ and $c^{\dagger}_{\mathbf{q}'\gamma}c_{\mathbf{q}\beta} = \delta_{\mathbf{q}\mathbf{q}'}\delta_{\beta\gamma}n_{\mathbf{q}}$, which are their expectation values with respect to the ground state $|\Omega\rangle$. We have also replaced H_c in the resolvent by its ground state expectation value, which is now zero. To simplify Eq. (11.29) we drop E, $\epsilon_{\mathbf{k}}$, $\epsilon_{\mathbf{k}'}$, whose magnitudes are small compared to D. Thus we have

$$H'^{(2)}_{\rm ex} = -\frac{|\delta D|\nu(0)}{2VD}\sum_{\mathbf{k}\mathbf{k}',\alpha\delta,ij} J_iJ_j(\sigma^i\sigma^j)_{\alpha\delta}(S_iS_j - S_jS_i)c^{\dagger}_{\mathbf{k}\alpha}c_{\mathbf{k}'\delta}. \tag{11.30}$$

Let us examine the second order correction proportional to $\sigma^z S^z$ in Eq. (11.30). We get

$$\frac{|\delta D|\nu(0)}{VD}J_{\perp}^2\sum_{\mathbf{k}\mathbf{k}',\alpha\delta}\sigma^z_{\alpha\delta}S^z c^{\dagger}_{\mathbf{k}\alpha}c_{\mathbf{k}'\delta}.$$

Likewise the correction to the transverse term is

$$\frac{|\delta D|\nu(0)}{VD}J_{\perp}J_z\sum_{\mathbf{k}\mathbf{k}'}\left(S^+c^{\dagger}_{\mathbf{k}\downarrow}c_{\mathbf{k}'\uparrow} + S^-c^{\dagger}_{\mathbf{k}\uparrow}c_{\mathbf{k}'\downarrow}\right).$$

Both these corrections are summarized in the following two coupled differential equations:

$$\begin{aligned}\frac{dJ_z}{d\ln D} &= -J_{\perp}^2\nu(0),\\ \frac{dJ_{\perp}}{d\ln D} &= -J_zJ_{\perp}\nu(0).\end{aligned} \tag{11.31}$$

Dividing the two equations above and integrating gives us the (hyperbolae) solutions

$$J_z^2 - J_{\perp}^2 = C, \tag{11.32}$$

where C is a constant. Fig.11.2 shows the scaling trajectories for this anisotropic Kondo model. Consider now the isotropic case with antiferromagnetic coupling $J > 0$. Integration of the scaling equations yields

$$J(D') = \frac{J(D)}{1 - J(D)\nu(0)\ln(D/D')}. \tag{11.33}$$

In particular for bandwidths such that $D' \sim T$ and with D as the bandwidth in the original high-energy model, Eq. (11.33) gives us the temperature dependence of the effective coupling constant $J(T)$. We also identify a single parameter for each scaling trajectory,

$$D\exp[-1/J(D)\nu(0)] = D'\exp[-1/J(D')\nu(0)] \equiv T_K,$$

that plays the role of an invariant with respect to scaling. Physically, systems lying on a scaling trajectory but with different values of D and J have the

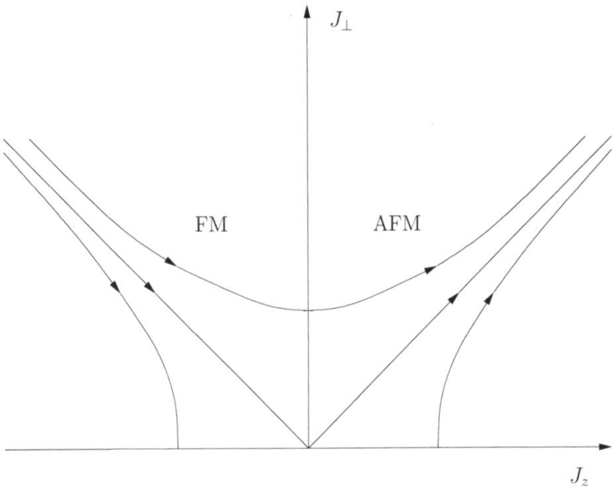

FIGURE 11.2: Solutions of the scaling equations (see Eq. (11.31)) of the anisotropic Kondo model. The arrows denote evolution in the direction of decreasing bandwidth D. FM and AFM denote ferromagnetic and antiferromagnetic Kondo effects respectively.

same low energy properties. Note that the temperature-dependent coupling constant $J(T)$ diverges as $T \to T_K$.

The results we obtained in Sec. 11.2 for impurity resistivity and susceptibility can be summarized in a concise way in terms of the effective coupling constant:

$$\begin{aligned} \rho_{\text{imp}} &= c_{\text{imp}} \frac{3\pi m \nu(0) J(T)^2}{2ne^2\hbar}, \\ \chi_{\text{imp}} &= c_{\text{imp}} (\mu_B^2/T) \left[1 - J(T)\nu(0)\right]. \end{aligned} \tag{11.34}$$

What happens to the impurity resistivity, susceptibility and other properties when $T < T_K$ where our perturbation theory is no longer valid? We have looked at some of the properties in this regime earlier in Chapter 5 by assuming a Fermi liquid ground state. But phenomenology by itself is not sufficient to answer whether the ground state and low energy excitations are described by a Fermi liquid theory. Wilson developed a numerical renormalization group technique that he used to obtain the properties of the Kondo model over the entire range of energies, including the ground state. It was then conclusively established that the ground state and low energy excitations of the Kondo model are indeed governed by a Fermi liquid fixed point (noninteracting theory with irrelevant interactions).

Having understood the scaling technique for the single-channel Kondo model, we now proceed to apply it to the multichannel Kondo model.

11.4 Nozières-Blandin theory of the overscreened Kondo model

A seemingly innocuous step of allowing the impurity spin to couple equally to electrons in more than one band leads to dramatic changes in the physical outcome. Let us consider the following variant of the isotropic Kondo model,

$$H = H_c + H_K \equiv \sum_{\mathbf{k}a\alpha} \epsilon_{\mathbf{k}} c^\dagger_{\mathbf{k}a\alpha} c_{\mathbf{k}a\alpha} + \frac{J}{V} \sum_{\mathbf{k}\mathbf{k}',\alpha\beta,a} c^\dagger_{\mathbf{k}a\alpha} \boldsymbol{\sigma}_{\alpha\beta} c_{\mathbf{k}a\beta} \cdot \mathbf{S}, \qquad (11.35)$$

where $a = 1, 2, \ldots, N_{\rm ch}$ is an additional label on the conduction electrons. This could arise, for example, if conduction electrons from a number of atomic orbitals coupled to the magnetic impurity. It turns out that the properties of this model can be very different from the Kondo model we have been discussing so far. The key understanding of multichannel Kondo models was developed by Nozières and Blandin[8]. We are interested in the $T-$matrix element,

$$T_{\mathbf{k}a\sigma,\mathbf{k}'a'\sigma'} = \langle \Omega | c_{\mathbf{k}a\sigma} T c^\dagger_{\mathbf{k}'a'\sigma'} | \Omega \rangle$$

for small couplings $J\nu(0) \ll 1$ and large channel number $N_{\rm ch} \gg 1$. The leading order contribution is easily shown to be

$$T^{(1)}_{\mathbf{k}a\sigma,\mathbf{k}'a'\sigma'} = \frac{J}{V} \delta_{aa'} (1 - n_{\mathbf{k}})(1 - n_{\mathbf{k}'}) \mathbf{S} \cdot \boldsymbol{\sigma}_{\sigma\sigma'}. \qquad (11.36)$$

Evaluation of the second order contribution is similar to the single-channel case we studied in the last section:

$$T^{(2)}_{\mathbf{k}a\sigma,\mathbf{k}'a'\sigma'} = (J^2 \nu(0)/V) \delta_{aa'} (1 - n_{\mathbf{k}})(1 - n_{\mathbf{k}'}) \ln \left| \frac{D}{E} \right| (\mathbf{S} \cdot \boldsymbol{\sigma})_{\sigma\sigma'}. \qquad (11.37)$$

We have used the new definition of E, replacing $(E - E_0) - \epsilon_F$ by E. Clearly, the contributions up to second order do not depend on $N_{\rm ch}$.

Consider now the third order term,

$$\begin{aligned} T^{(3)}_{\mathbf{k}a\sigma,\mathbf{k}'a'\sigma'} &= \frac{J^3}{V^3} \sum_{\text{ind. labels}} \langle \Omega | c_{\mathbf{k}a\sigma} c^\dagger_{\mathbf{q}b\alpha} c_{\mathbf{q}'b\alpha'} \frac{1}{E - H_c} c^\dagger_{\mathbf{p}c\beta} c_{\mathbf{p}'c\beta'} \\ &\quad \times \frac{1}{E - H_c} c^\dagger_{\mathbf{r}d\gamma} c_{\mathbf{r}'d\gamma'} c^\dagger_{\mathbf{k}'a'\sigma'} | \Omega \rangle (\mathbf{S} \cdot \boldsymbol{\sigma})_{\alpha\alpha'} (\mathbf{S} \cdot \boldsymbol{\sigma})_{\beta\beta'} (\mathbf{S} \cdot \boldsymbol{\sigma})_{\gamma\gamma'} \\ &= \frac{J^3}{V^3} \sum_{\text{ind. labels}} \frac{\langle c_{\mathbf{k}a\sigma} c^\dagger_{\mathbf{q}b\alpha} c_{\mathbf{q}'b\alpha'} c^\dagger_{\mathbf{p}c\beta} c_{\mathbf{p}'c\beta'} c^\dagger_{\mathbf{r}d\gamma} c_{\mathbf{r}'d\gamma'} c^\dagger_{\mathbf{k}'a'\sigma'} \rangle}{[E - (\epsilon_{\mathbf{q}'} - \epsilon_{\mathbf{q}} + \epsilon_{\mathbf{k}})][E - (\epsilon_{\mathbf{r}} - \epsilon_{\mathbf{r}'} + \epsilon_{\mathbf{k}'})]} \\ &\quad \times (\mathbf{S} \cdot \boldsymbol{\sigma})_{\alpha\alpha'} (\mathbf{S} \cdot \boldsymbol{\sigma})_{\beta\beta'} (\mathbf{S} \cdot \boldsymbol{\sigma})_{\gamma\gamma'}. \end{aligned} \qquad (11.38)$$

[8]P. Nozières and A. Blandin, *J. Physique* **41**, 193 (1980).

Here the summations are to be carried out over all internal independent degrees of freedom (referred to as "ind. labels"). The third order contributions are smaller than the second order ones by an extra factor of $J\nu(0)$; however, there is one new kind of contribution (absent in lower orders) that is proportional to $N_{\rm ch}$, which arises as follows. In Eq. (11.38) we make the contractions $\langle c_{\mathbf{k}a\sigma}c^\dagger_{\mathbf{p}c\beta}\rangle$ and $\langle c_{\mathbf{p}'c\beta'}c^\dagger_{\mathbf{k}'a'\sigma'}\rangle$. The remaining contractions $\langle c^\dagger_{\mathbf{q}b\alpha}c_{\mathbf{r}'d\gamma'}\rangle\langle c_{\mathbf{q}'b\alpha'}c^\dagger_{\mathbf{r}d\gamma}\rangle$ involve the sum $\sum_{bd}\delta_{bd} = N_{\rm ch}$. This contraction results in a fermion "bubble" and summing over the internal channel index in this bubble gives the factor $N_{\rm ch}$. We specifically consider this term and neglect the rest of the third order terms since $N_{\rm ch} \gg 1$. So far we have

$$T^{(3)}_{\mathbf{k}a\sigma,\mathbf{k}'a'\sigma'} \approx \frac{J^3 N_{\rm ch}}{V^3}\delta_{aa'}(1-n_{\mathbf{k}})(1-n_{\mathbf{k}'}) \times \sum_{\text{ind. labels}} \frac{n_{\mathbf{q}}(1-n_{\mathbf{q}'})(\mathbf{S}\cdot\boldsymbol{\sigma})_{\alpha\alpha'}(\mathbf{S}\cdot\boldsymbol{\sigma})_{\sigma\sigma'}(\mathbf{S}\cdot\boldsymbol{\sigma})_{\alpha'\alpha}}{[E-(\epsilon_{\mathbf{q}'}-\epsilon_{\mathbf{q}}+\epsilon_{\mathbf{k}})][E-(\epsilon_{\mathbf{q}'}-\epsilon_{\mathbf{q}}+\epsilon_{\mathbf{k}'})]}. \tag{11.39}$$

The trace over the spins can be simplified:

$$\sum_{ijk,\alpha} S_i S_j S_k \sigma^j_{\sigma\sigma'}(\sigma^i\sigma^k)_{\alpha\alpha} = 2(S_x^2 - S_y^2 - S_z^2)(S_x\sigma_x)_{\sigma\sigma'} + \text{permutations.}$$

For the special case of $S = 1/2$, the trace over spins gives us $-(1/2)(\mathbf{S}\cdot\boldsymbol{\sigma})_{\sigma\sigma'}$. Incorporating this in Eq. (11.39) we get

$$\begin{aligned} T^{(3)}_{\mathbf{k}a\sigma,\mathbf{k}'a'\sigma'} &\approx -\frac{J^3 N_{\rm ch}}{2V}\delta_{aa'}(1-n_{\mathbf{k}})(1-n_{\mathbf{k}'})(\mathbf{S}\cdot\boldsymbol{\sigma})_{\sigma\sigma'}\sum_{\mathbf{q}\mathbf{q}'}\frac{n_{\mathbf{q}}(1-n_{\mathbf{q}'})}{[E-(\epsilon_{\mathbf{q}'}-\epsilon_{\mathbf{q}})]^2} \\ &= -\frac{J^3\nu(0)^2 N_{\rm ch}}{8V}\delta_{aa'}(1-n_{\mathbf{k}})(1-n_{\mathbf{k}'})\ln|D/E|(\mathbf{S}\cdot\boldsymbol{\sigma})_{\sigma\sigma'}. \end{aligned} \tag{11.40}$$

We can thus infer the following poor man's scaling law for the multichannel Kondo model,

$$\frac{dJ}{d\ln D} = -J^2\nu(0) + \frac{N_{\rm ch}}{8}J^3\nu(0)^2. \tag{11.41}$$

Eq. (11.41) has a stable fixed point at $J_*\nu(0) = 8/N_{\rm ch}$, and the Kondo temperature is exponentially small in the number of channels, $T_K \sim De^{-1/J_*\nu(0)} = De^{-N_{\rm ch}/8}$. If $N_{\rm ch} \gg 8$, the fixed point lies within the regime of validity of our perturbative analysis. To see this, observe that at the fixed point, each interaction vertex J_* comes with a factor $1/N_{\rm ch}$, and therefore each fermion loop comes with a factor $J_*^2 N_{\rm ch} \sim 1/N_{\rm ch}$. Thus in Eq. (11.41), we drop higher order terms in J on the RHS. Clearly, for small values of $N_{\rm ch}$, many more terms come into play and any nontrivial fixed point that is not at zero or infinity may not be accessible by a perturbative calculation.

We conclude with a discussion of the stability of fixed points for a general problem of a spin-S impurity isotropically Kondo-coupled to $N_{\rm ch}$ channels of conduction electrons. Nozières and Blandin showed that the stability depends on the ratio $2S/N_{\rm ch}$. The strong coupling limit ($J\nu(0) \to \infty$) corresponds to an impurity spin of size $S' = S - N_{\rm ch}/2$ interacting with the conduction electrons. When $2S/N_{\rm ch} > 1$, the orientation of the dressed spin given by $S'_z = S_z - N_{\rm ch}/2$ is in the same direction as S_z. This is referred to as the *underscreened* Kondo model. In the underscreened case, the interaction between S'_z and the conduction electrons is of the order of D^2/J from second order perturbation theory. This interaction is *ferromagnetic* as is seen by the following simple argument. Suppose S_z is along the positive z direction, and is screened by $N_{\rm ch}$ conduction electrons with down spin. Any conduction electron that makes a virtual hop to this site must have an up-spin because Pauli's principle forbids the hopping of any more down spins (these states are all occupied). Thus the interaction is ferromagnetic. Now if we do a poor man's scaling from this end for the ferromagnetic Kondo model, the flow would be back towards the strong coupling fixed point. Thus $J = \infty$ is a stable fixed point in this case. When $2S/N_{\rm ch} < 1$, S'_z points in the opposite direction to S_z, and the Kondo interaction between S'_z and the conduction electrons is now *antiferromagnetic.* That makes the $J = \infty$ fixed point unstable and the flow is back towards weak coupling. In that case, there must be a stable fixed point at intermediate coupling. This is the *overscreened* Kondo model. The $N_{\rm ch} = 2S$ is special since it corresponds to $S' = 0$. In this case, the ground state consists of a many-body spin singlet formed by the impurity and the conduction electrons. The physics at low but finite temperatures is described by Nozières' Fermi liquid theory for the Kondo problem, which we studied in Chapter 5 for the case $N_{\rm ch} = 2S = 1$. An important outcome of the Fermi liquid treatment was the value of the Wilson ratio, $R = \frac{(\chi_{\rm imp}/\chi)}{(C_{\rm imp}/C)} = 2$. Following essentially the same arguments, one can show that for an isotropic model with $N_{\rm ch}$ channels, the Wilson ratio is $R = \frac{2}{3}(N_{\rm ch} + 2)$. The impurity specific heat coefficient also diverges logarithmically as $T \to 0$, in marked contrast to the Fermi liquid fixed point behavior, where the specific heat coefficient approaches a constant value.

Poor man's scaling is not adequate for obtaining impurity properties in the low temperature regime of overscreened Kondo models. A number of other methods are available, ranging from numerical renormalization group to bosonization and Bethe ansatz techniques. For the special case of the two-channel Kondo model, $S = \frac{1}{2}$ and $N_{\rm ch} = 2$, the impurity entropy turns out to be $\frac{1}{2}\ln 2$, i.e., half the value expected for a free spin[9].

11.4.1 Magnetic impurities in almost ferromagnetic metals

Nearly a decade before the study of the multichannel Kondo model by Nozières and Blandin, Larkin and Mel'nikov in 1971 serendipitously discov-

[9] P. Coleman, L. B. Ioffe and A. M. Tsvelik, *Phys. Rev. B* **52**, 6611 (1995).

ered[10] the multichannel Kondo effect while studying the effect of a magnetic impurity in the vicinity of a ferromagnetic quantum critical point, although they did not recognize the phenomenon in its entirety. This was pointed out by the author and co-workers[11] in 2005.

The basic motivation for Larkin-Mel'nikov theory is quite simple. Near a ferromagnetic critical point, the magnetic susceptibility tends to diverge. Magnetic impurities introduced into the metallic host in this regime polarize a large number of conduction electrons, forming a droplet with a giant magnetic moment. Fluctuations of this droplet are governed by the dynamic magnetic fluctuations of the metallic host. We know that such fluctuations are overdamped. The question is whether this overdamping completely suppresses impurity spin-flip scattering, and by implication, the Kondo effect. We present a sketch of the Larkin-Mel'nikov treatment.

The conduction electrons couple to the impurity spin through the usual Kondo coupling:

$$H_{\mathrm{ex}} = JV \sum_{\alpha\beta} \int_{\mathbf{k},\mathbf{k}'} c^{\dagger}_{\mathbf{k}\alpha} \boldsymbol{\sigma}_{\alpha\beta} c_{\mathbf{k}'\beta} \cdot \mathbf{S}.$$

The difference from the usual story is that now the conduction electrons are on the verge of ferromagnetic order. Approaching a ferromagnetic quantum critical point (QCP) from within the "quantum" (or Landau Fermi liquid) regime of Hertz theory, the correlation function of magnetic fluctuations is,

$$D(i\omega_n, \mathbf{q}) = \frac{1}{r + K\mathbf{q}^2 + \frac{|\omega_n|}{\gamma|\mathbf{q}|}}, \tag{11.42}$$

where $r = 0$ is the critical point, K the spin-wave stiffness and $\gamma = (2/\pi)v_F$ the Landau damping coefficient. The static (Pauli) susceptibility, $\chi = \nu(0)\mu_B^2 D(0, \mathbf{q} \to 0) = \nu(0)\mu_B^2/r$, diverges upon approaching the QCP.

Next one performs a scaling analysis of the interaction vertex J. The leading order in perturbation theory (i.e. J^2) gives us the usual Kondo effect. As we have seen in our study of the multichannel Kondo models, the J^3 term involves a particle-hole bubble. Although we do not formally have multiple channels[12] here, the presence of a particle-hole bubble amplifies this contribution by a factor $1/r$. Sufficiently close to the critical point, the third order contribution may even dominate the leading order term; that happens when $\frac{1}{r}(J\nu(0)) \gg 1$. Larkin and Mel'nikov focused on this regime. They first traced over the conduction electron degrees of freedom to obtain the reduced density

[10] A. I. Larkin and V. I. Mel'nikov, *Sov. Phys. JETP* **34**, 656 (1972) [*Zh. Eksp. Teor. Fiz.* **61**, 1231 (1971)]

[11] Y. L. Loh, V. Tripathi and M. Turlakov, *Phys. Rev. B* **71**, 024429 (2005).

[12] One can physically reason that there are effectively multiple channels here, see following analysis.

matrix for the impurity spin:

$$\rho_{\text{imp}}(\beta) \approx \frac{1}{\mathcal{Z}} T_\tau \exp\left(-\frac{J^2}{2} \int_0^\beta d\tau \int_0^\beta d\tau' \, \mathbf{S}(\tau) \cdot \mathbf{S}(\tau') L(\tau - \tau') \right), \tag{11.43}$$

where $L(\tau - \tau') = \frac{1}{\beta} \sum_n e^{-i\omega_n(\tau-\tau')} L(i\omega_n)$ is the (Matsubara) Fourier transform of the *local* correlation function of the magnetization:

$$L(i\omega_n) - L(0) = \nu(0) \int (d\mathbf{q})[D(i\omega_n, \mathbf{q}) - D(0, \mathbf{q})]$$
$$\approx \begin{cases} -\frac{A\pi\nu(0)^2}{2r}|\omega_n|, & |\omega_n| \ll E_{\text{C}} = \epsilon_F r^{3/2} \\ -B\nu(0)^2 \epsilon_F \left(\frac{|\omega_n|}{\epsilon_F}\right)^{1/3}, & E_{\text{C}} \ll |\omega_n| \ll \epsilon_F, \end{cases}$$

and A, B are constants of order unity whose details are not important for us. The $\mathbf{S}(\tau)$ are spin-operators satisfying the usual angular momentum algebra and T_τ denotes time ordering. Since we are interested in excitation far smaller than the Fermi energy, we may approximately regard the energy scale $E_{\text{C}} = \epsilon_F r^{3/2} \equiv \epsilon_F \xi^{-z}$ as a cutoff for $|\omega_n|$. At low energies, the effective interaction in Eq. (11.43) has the form $J^2 L(i\omega_n) = J^2\nu(0)L(0) - g(\pi/2)|\omega_n|$, with the coupling g given by

$$g = A(J\nu(0))^2 \frac{1}{r}. \tag{11.44}$$

A perturbative treatment in increasing powers of g is possible provided $g < 1$. For this, it is necessary not to begin too close to the QCP. Using perturbative RG, Larkin and Mel'nikov calculated the scaling of the coupling constant. Denoting $\omega = E_{\text{C}} e^{-\ell}$, the scaling of the interaction vertex goes as

$$g(\ell) = \frac{g}{1 + g\ell} \equiv \frac{g}{1 + g \ln |E_{\text{C}}/\omega|}. \tag{11.45}$$

Thus the coupling $g = A(J\nu(0))^2/r$ is marginally irrelevant and flows towards zero upon scaling. Re-expressing the scaling in Eq. (11.45) in terms of J and introducing back the leading order Kondo scaling term one has

$$\frac{dJ\nu(0)}{d\ell} = +(J\nu(0))^2 - 2A(J\nu(0))^3/r. \tag{11.46}$$

We immediately recognize Eq. (11.46) as the scaling equation for the interaction vertex in the overscreened Kondo model with the channel number $N_{\text{ch}} \sim 1/r$.. A nonzero stable fixed point exists at $J_*\nu(0) = r/2A$, with a Kondo temperature $T_K \sim E_{\text{C}} e^{-2A/r}$ that vanishes on approaching the QCP.

What happens when the initial conditions are such that $g > 1$? That always happens sufficiently close to the QCP. One possibility is that fluctuations of the impurity spin get frozen due to the strong overdamping, cutting off the Kondo effect. The other possibility is that the critical fluctuations of the conduction electron spin density lead to a renormalization of the coupling g towards small values whereupon the usual Larkin-Mel'nikov machinery takes over. It turns out that the latter possibility is the correct one[13]. Beginning with a large coupling $g \gg 1$, it has been shown that upon scaling, the coupling renormalizes to smaller values in accordance with the scaling equation

$$g(\ell) = g - \frac{1}{\pi}\ell + O(g^2), \tag{11.47}$$

with $\ell \sim \ln(gE_{\rm C}/\omega)$. Observe that there is no first order in g term on the RHS. At a sufficiently small energy scale ω_* such that $\omega_* \sim E_{\rm C}e^{-\pi g}$, the effective coupling becomes order one, and one may continue scaling using the Larkin-Mel'nikov prescription until the multichannel Kondo fixed point is reached.

There are several experimental systems, involving impurities with giant magnetic moments in a nearly ferromagnetic host metal, in which it may be possible to study magnetic droplet phenomena systematically. Among these are iron dissolved in various transition metal alloys, nickel impurities in palladium and cobalt impurities in a platinum host.

Kondo effects generically arise whenever magnetic impurities interact with low-lying magnetic excitations in a paramagnetic host. Recent examples of such systems include the zinc-doped cuprates in their putative spin-liquid phase, and magnetic impurities in the honeycomb Kitaev model. Scaling analysis may be performed in such instances, in regimes where a small parameter is available for the perturbative treatment. In particular, for the magnetically-doped Kitaev model in its gapped spin-liquid phase, a very unusual Kondo effect results[14,15], with an *unstable* fixed point separating strong and weak coupling regimes.

[13] Y. L. Loh, V. Tripathi and M. Turlakov, *Phys. Rev. B* **71**, 024429 (2005).
[14] K. Dhochak, R. Shankar and V. Tripathi, *Phys. Rev. Lett.* **105**, 117201 (2010).
[15] S. D. Das, K. Dhochak and V. Tripathi, *Phys. Rev. B* **94**, 024411 (2016).

Part IV

Open Questions

Chapter 12

Electron Lifetime in a Mesoscopic Conductor

Analyzing disorder effects in three-dimensional Landau Fermi liquids in Chapter 8, we showed that the quasiparticle relaxation rate[1] $1/\tau_{\rm ee} \sim |\epsilon|^{3/2}$, is larger than in the clean Fermi liquid in which $1/\tau_{\rm ee} \sim \epsilon^2$. The calculation of the relaxation rate was based on application of the Fermi Golden Rule, which is relevant where the final states in a scattering process can be regarded as a continuum. Coulomb interaction provides the means by which a particle scatters into another particle state together with a particle-hole pair. Physically, the enhancement of the decay rate in the disordered case arises from reduction of some of the phase space constraints related to momentum conservation, which in the clean Fermi liquid lead to the ϵ^2 dependence. We pose the following question in this chapter: what happens to the electron relaxation rate in a disordered metal when its volume is progressively decreased? The point is that in a finite volume, the energy levels become discrete and one needs to exercise care while using the Fermi Golden Rule, or for that matter, even understanding the meaning of quasiparticle relaxation.

12.1 Meaning of quasiparticle decay

The first task we face is how to describe a quasiparticle decay process, since all eigenstates of interacting particles in a finite volume, with generically discrete eigenvalues, are supposed to be infinitely long-lived. While any one of the many-particle eigenstates is indeed infinitely long-lived, we could ask whether and to what extent the many-particle eigenstates overlap with a given single particle excitation that may, for instance, be created in an actual experiment. A number of many-particle states could have a fair amount of

[1] All excitation energies are to be reckoned from the chemical potential.

overlap with this single-particle excitation, and the spread of energies of the many-particle states is a measure of the "decay rate" of the single-particle excitation. Suppose a single particle excitation $|\psi\rangle$ is created in the interacting system at a time $t = 0$, and the excitation overlaps appreciably with a set of eigenstates $|\{i\}\rangle$ of the many-particle system, with corresponding eigenvalues E_i. At a later time t, the wavefunction of the single-particle excitation will have the form

$$|\psi(t)\rangle = \sum_i a_i e^{-iE_i t}|i\rangle. \tag{12.1}$$

Suppose the energies E_i in this linear superposition have a finite spread, ΔE. Then after a time $\Delta t \sim 2\pi/\Delta E$, the original phase relation of the different components would be lost, and the quasiparticle can be regarded as having decayed. Thus we say $\tau_{\mathrm{ee}}^{-1} \sim \Delta E$. For a well-defined quasiparticle, one requires this spread to be smaller than the quasiparticle energy itself. This is not really a completely new definition of the quasiparticle decay rate. In the conventional approach, a single particle excitation, not being an eigenstate, is said to decay through the creation of a particle-hole shower or many-particle excitations, and the decay is to a set of exact many-particle eigenstates whose energies are close to that of the single particle excitation. The Golden Rule simply relates the decay rate to the overlap of the single-particle excitation with multi-particle ones. When the spread of energies of many-particle states that appreciably overlap with the single-particle excitation becomes comparable to the single-particle excitation energy, we say that quasiparticles are no longer well-defined excitations. A commonly encountered situation where this generically happens is an interacting one-dimensional conductor, where single-particle excitations are not well-defined.

Consider now the finite volume case. For single-particle excitations at energies large compared to the (single-particle) mean-level spacing δ, we can regard the energy levels as a continuum (density of final shower of quasiparticles is large) and the Fermi Golden Rule can be expected to apply. But for energies that are comparable to δ, Golden Rule considerations no longer apply, and we must revert to the more accurate definition of the quasiparticle decay rate as a spread of energies of the many-particle eigenstates with which the single-particle excitation has an appreciable overlap. In fact, when the density of final states is no longer large, Golden Rule considerations merely give perturbative corrections to the energies of single-particle states. When the single-particle energy approaches δ, then one expects only a small number of many-particle states with which the single-particle excitation has a significant overlap. The implication is that the quasiparticle does not decay since it will take a very long time before the phases of the small number of closely spaced many-particle eigenstates lose their original phase relations. The extreme limits $\epsilon \gg \delta$ and $\epsilon \simeq \delta$ are thus expected to behave very differently. However, to find the characteristic energy scale below which the decay rate becomes qualitatively small compared to the bulk value, we need to look at the problem more closely. Following a careful analysis of this problem[2] a surprising suggestion was made that the electron decay rate *completely* vanishes at a transition energy ϵ_T that is still parametrically large compared to the mean level spacing.

12.2 Qualitative estimate of transition energy

We consider a diffusive metallic grain of size L, electron diffusion constant D and mean level spacing δ. The metallic grain is characterized by a dimensionless conductance $g = E_T/\delta \gg 1$, i.e., the ratio of the Thouless energy $E_T = D/L^2$ and the mean level spacing. The electrons in the grain interact via short-range four fermion interactions which can also be regarded as the matrix element between a single particle state and a three-particle state consisting of two particles and one hole (the four fermion interaction has two creation and two annihilation operators). Our Hamiltonian is

$$H = H^{(0)} + H^{(1)} = \sum_{\alpha} \epsilon_\alpha c_\alpha^\dagger c_\alpha + \sum_{\alpha\beta\gamma\delta} U_{\gamma\delta}^{\alpha\beta} c_\gamma^\dagger c_\delta^\dagger c\beta c_\alpha. \tag{12.2}$$

Here α labels the single particle eigenstates of the disordered conductor and the single-particle energies ϵ_α take random values from some distribution whose details are not important for the present discussion. We consider spinless electrons for simplicity. $U_{\gamma\delta}^{\alpha\beta} = \int_{\mathbf{r},\mathbf{r}'} \psi_\delta^*(\mathbf{r})\psi_\gamma^*(\mathbf{r}')U(\mathbf{r}-\mathbf{r}')\psi_\beta(\mathbf{r})\psi_\alpha(\mathbf{r}')$ is the matrix element of the short-range Coulomb interaction,

$$U(\mathbf{r}-\mathbf{r}') = V\lambda\delta\,\delta(\mathbf{r}-\mathbf{r}'), \tag{12.3}$$

where V is the volume and the dimensionless strength $\lambda \sim 1$. In the random system, $U_{\gamma\delta}^{\alpha\beta}$ is a fluctuating quantity with zero mean and we characterize it by the root mean square (RMS) value in an energy window of size E_T. Averaging $(U_{\gamma\delta}^{\alpha\beta})^2$ over disorder gives us[3] an RMS value $U \sim \lambda\delta/g$. One may understand this as follows. We interpret the Coulomb interaction vertex (of strength $\lambda\delta$) as representing the scattering of a low-energy particle-hole (diffuson) excitation and the propagator for these density fluctuations, as we saw in Chapter 8, is $1/(D\mathbf{q}^2 - i\omega)$. The matrix element $(U_{\gamma\delta}^{\alpha\beta})^2$ consists of a diffuson loop with two Coulomb vertices. We sum over frequencies ω in a window of size E_T, and obtain the estimate $U^2 \sim (\lambda\delta/g)^2$

When a quasiparticle with energy $\epsilon \gg \delta$ decays into a shower of two particles and one hole (subject to conservation of energy), the final products have

---

[2]See B. L. Altshuler, Y. Gefen, A. Kamenev and L. S. Levitov, *Phys. Rev. Lett.* **78**, 2803 (1997). We shall broadly take the approach of this paper.

[3]See reference in Footnote 2.

access to a large number, $N_3 \sim (\epsilon/\delta)^3$, of states, which translates to a density of states for the three-particle shower, $\nu_3 = \partial N_3/\partial\epsilon \sim \epsilon^2/\delta^3$. More accurately, imagine we have $N = (\epsilon/\delta) \gg 1$ objects that we need to arrange in three boxes, the number of objects in the three boxes (separated by two partitions) representing the energies of the particles. The number of ways the distribution can be done is $(N+2)!/(N!2!)$, since we can permute the objects and the partitions among themselves without generating a new arrangement. For $N \gg 1$, we get $\nu_3 \sim \frac{1}{\delta}(\epsilon/\delta)^2/2$. The validity of the Golden Rule calculation for electron decay rate rests on the quantity $U\nu_3 = \epsilon^2/(g\delta^2)$ far exceeding unity, for that allows the matrix element to connect the initial quasiparticle state to a large number of final three-particle states. When $U\nu_3$ becomes small compared to unity, then the Golden Rule merely results in perturbative corrections to the energy levels but no decay. The energy scale at which this transition takes place is thus $\epsilon^* \sim \delta\sqrt{g} \gg \delta$. The question arises if this represents a crossover or whether the quasiparticle decay rate completely vanishes below some energy scale of the order of ϵ^*. We present arguments that the latter is true.

12.3 Localization transition in Fock space

We now re-interpret the interaction term in Eq. 12.2 as a two-particle hopping in the (Fock) space of states $\{\alpha\}$ involving annihilation of a fermion pair and creation of another one. We begin with the ground state of $(N-1)$ particles in the grain and add a single particle of energy ϵ to the system. We call this a generation-1 process. In generation-3 that comes next, the interaction term annihilates this particle and creates two particles and one hole, subject to the constraint that the total energy is conserved. The density of final states in this shower is $\nu_3 \sim \frac{1}{2\delta}(\epsilon/\delta)^2$. In this manner we construct successive $(2n+1)$-generation showers. The density of states of the final products in generation-$(2n+1)$ is $\nu_{2n+1} = \frac{1}{\delta}(N+2n)!/(N!(2n)!) \sim (1/\delta)(\epsilon/\delta)^{2n}/(2n)!$, where the factorial comes from permuting the $2n$ partitions that segregate $N = (\epsilon/\delta)$ objects into $(2n+1)$ boxes. We want a related but slightly different quantity. We would like to know the density of states associated with a three-particle shower coming from an individual arm of the previous generation-$(2n-1)$. This number is $(\nu_{2n+1})^{1/(2n)} \sim \nu_3/(2n)$. The above branching processes are not the only ones that we have. The interaction term can also annihilate a particle-hole pair in generation-$(2n-1)$ (instead of creating an extra pair), and take us to the previous one. This is overall equivalent to selecting two partitions from the original $(2n-2)$ and removing them. The number of ways is ${}^{2n-2}C_2 = (n-1)(2n-3)$, and corresponding density of states $(n-1)(2n-3)/\delta$. It is also possible to annihilate a pair of particles in generation-$(2n-1)$ and create another pair elsewhere – this process clearly does not change the generation.

Such a process is equivalent to swapping any of the N objects with any of the $(2n-2)$ partitions – the number of possibilities is clearly $(2n-2)N$, and hence the density of final states for this process is $\frac{1}{2\delta}(n-1)N \sim (n-1)\sqrt{\nu_3/\delta}$. For large values of N, it is evident that the maximum density of final states is associated with the branching process. We will henceforth disregard the ones involving motion within a generation or to the previous one. We then have a tree like structure, with the difference that the branching in generation-$(2n+1)$ is not constant but is given by

$$K_{2n+1} \sim \frac{1}{n}\int_0^{E_T} \nu_3(\epsilon)\, d\epsilon = \frac{1}{6n}g^3, \tag{12.4}$$

which gradually falls with increasing n.

The problem of quasiparticle decay has thus been mapped on to a problem of a free particle hopping along the arms connecting successive vertices on the above graph (an object in the Fock space of states), and subjected to a random potential[4]. We would like to know if the particle executing motion on the tree can get localized (i.e., not reach the bottom of the tree) for sufficiently strong disorder. If it does get localized, then it is analogous to saying the injected particle creates at the most some few-particle excitations, and is thus a well-defined excitation since there is appreciable overlap between the few-particle state and the single-particle state. On the other hand, if the disorder is sufficiently weak, the motion in Fock space is delocalized. If the initial single-particle state overlaps with a large number of many-particle states (trajectories in Fock space), we say the quasiparticle is not a well-defined excitation. A localization-delocalization transition of a quantum particle subjected to a random potential is known in the literature as Anderson localization.

When the branching ratio K is independent of n, the graph is known as a Cayley tree. We also make the drastic approximation here that in every generation, $K = g^3/6$ and does not decrease with n. This simplifies our further analysis enormously. The problem of localization of a quantum particle on a Cayley tree by disorder has been studied earlier in the context of Anderson localization. Denoting the hopping element by a constant, t, and the width of the disorder distribution by W, it can be shown that the transition occurs at[5]

$$\frac{W}{t} = K \ln K, \tag{12.5}$$

where K is the branching ratio of the tree. We also make a similar drastic approximation for the density of final states in a generation-$(2n+1)$ shower associated with a particular arm of generation-$(2n-1)$. Instead of the correct result ν_3/n, we use the n-independent value ν_3. The interaction strength $U \sim \lambda\delta/g$ plays the role of hopping element t in Eq. (12.5). Finding what plays the

[4]The particle coordinate here refers to the state of the system and not the actual coordinates of the injected particle.

[5]See R. Abou-Chacra, P. W. Anderson, and D. J. Thouless, *J. Phys. C* **6**, 1734 (1973).

role of the energy spread W is a bit tricky. Now in our three-particle shower, the density of final states, $\nu_3 = N/\delta$, is much larger than $1/\delta$, and the mean spacing of these many-particle excitations is δ/N. But there are $K = g^3/6$ states in every shower, so the spread of energies in the three-particle shower is $W(\epsilon) \sim K\delta/N \sim K\delta/(\epsilon/\delta)^2$. The condition for Anderson transition is then satisfied at an energy ϵ_T satisfying $W(\epsilon_T)/U = K \ln K$, or,

$$\epsilon_T \sim \delta\sqrt{\frac{g}{\lambda \ln g}}, \tag{12.6}$$

an improvement on our earlier estimate $\epsilon_T \sim \delta\sqrt{g}$ based on Fermi Golden rule considerations. Note that there is no actual localization-delocalization transition in physical space. This transition, taking place in the Fock space, is an example of the so-called many-body localization phenomenon.

We do not know the answer to a number of questions. First, in retaining only the branching processes that advance the generations, we lost the effects coming from processes that did not change the generation or went back in generations. Although parametrically smaller, it is not clear how their presence affects the above conclusions. Second, we do not quite have an infinite Cayley tree because the branching extends only up to a maximum generation $n_{\max} \sim \epsilon/\delta$. On the finite tree, there is no real delocalization transition. Finally, we do not know how our results would change if the n-dependence of the branching ratio of our Cayley tree is not neglected. The analysis presented above is evidently geared towards making use of known results for Anderson localization on the Cayley tree.

Chapter 13

Marginal Fermi Liquids

The Landau Fermi liquid theory of metals has been so successful that exceptions to it draw a fair amount of attention. The theory rests on the assumption that the low-energy properties of interacting electrons in a metal in three dimensions may be described in terms of weakly repulsive charged spin-$\frac{1}{2}$ fermionic excitations – the quasiparticles. This is of course not generally expected to hold following a phase transition, for example, in the form of a Pomeranchuk instability or a superconducting transition. In Chapter 2 we showed that Pomeranchuk instabilities occur whenever some Landau F-function satisfies the condition $F_l^{s(a)}/(2l+1) = -1$, where $l = 0, 1, 2, \ldots$ denotes the relevant angular momentum channel where the instability occurs. The $l = 0$ instability in the spin-symmetric channel is associated with a divergence of compressibility that leads to phase separation and consequent breakdown of the Fermi liquid. In the spin-antisymmetric channel, the $l = 0$ instability corresponds to ferromagnetism. In the ferromagnetic quantum critical region ("classical" Gaussian regime in Hertz theory), we found in Sec. 9.6.2.3 that spin fluctuations lead to the destruction of the Landau Fermi liquid and this is evident, for example, in the $\ln(1/T)$ divergence of the specific heat coefficient $\gamma(T)$. Proximity to an electronic instability is however not a necessary condition for the breakdown of the Fermi liquid. In Sec. 10.2 we found that transverse electromagnetic fluctuations are only weakly screened by conduction electrons and that results in a $\gamma(T) \sim \ln(1/T)$ divergence for the specific heat coefficient without needing to be in the vicinity of a ferromagnetic quantum critical point[1]. The lack of screening of such electromagnetic fluctuations is a feature shared with low energy spin fluctuations in the Hertz theory of quantum critical ferromagnets. Another route to non Fermi-liquid behavior, not reliant on the existence of long-range interactions, appeared in Sec. 11.4 where we analyzed non-optimally screened Kondo models. It was found that singularities in properties such as the susceptibility and specific heat coefficient may be expected in the event that residual low-energy degeneracies associated with spin-flip scattering with conduction electrons persist down to the lowest temperatures. Finally, weak potential disorder was found to preserve the Fermi liquid (in three dimensions) down to the lowest temperatures (see Chapter 8).

[1] Although that would be a bonus.

The ingredient common to all the instances of the breakdown of the Fermi liquid we have come across in this book so far has been the availability of degenerate or a continuum of dynamic fluctuations with energies much smaller than the temperature, and interacting with the conduction electrons in some manner. The divergence of the specific heat coefficient $\gamma(T) = p_F m^*/3$ in all these cases implies a corresponding singularity in the quasiparticle self-energy which is related to the effective mass m^* and the quasiparticle weight $\mathcal{Z}$ (see Eqs. (9.8) and Eq. (9.10) and following discussion). If we interpret the divergence of $\gamma(T)$ from a Fermi liquid perspective, then $m^* \sim \ln(1/T)$, implying a vanishing of the quasiparticle weight, $\mathcal{Z} \sim 1/\ln(1/T)$. The phenomenologically proposed "marginal" self energy[2],

$$\Sigma(\epsilon, \mathbf{p}_F) \simeq \lambda[\epsilon \ln(|x|/\epsilon_c) + i\frac{\pi}{2}|x|], \tag{13.1}$$

$x = \max(|\epsilon|, T)$, gives $\mathcal{Z}^{-1} = 1 - \partial \mathrm{Re}\Sigma/\partial\epsilon \sim \ln(\epsilon_c/x)$, and is consistent with $\gamma(T) \sim \ln(1/T)$ at low temperatures. ϵ_c is a suitable high energy cutoff above which Eq. (13.1) is not expected to hold. The rationale for the name "marginal" is that Landau quasiparticles whose decay rate is determined by $\mathrm{Im}\Sigma \sim |\epsilon|^\alpha$, $\alpha = 2$, are no longer well-defined once $\alpha \leq 1$. The word marginal denotes the first power, $\alpha = 1$, where the quasiparticles just cease to remain well-defined excitations. The peculiar form of the real part of the self energy follows from the Kramers-Kronig equations,

$$\mathrm{Re}\Sigma(\epsilon) = \frac{1}{\pi} P \int_{-\infty}^{\infty} d\epsilon' \frac{\mathrm{Im}\Sigma(\epsilon)}{\epsilon' - \epsilon}, \tag{13.2}$$

$$\mathrm{Im}\Sigma(\epsilon) = -\frac{1}{\pi} P \int_{-\infty}^{\infty} d\epsilon' \frac{\mathrm{Re}\Sigma(\epsilon)}{\epsilon' - \epsilon}. \tag{13.3}$$

Here "P" denotes the principal value. In our discussion following the first Ward identity of landau Fermi liquids, Eq. (9.33), we had noted that the quasiparticle weight is a measure of the overlap of a single particle state and a quasiparticle state, with the latter involving one or more particle-hole pairs over and above the single particle excitation. The vanishing of $\mathcal{Z}$ in the marginal Fermi liquid tells us that in such a state, it is not possible to have well-defined single fermion excitations. The vanishing of the overlap of the single fermion state and the single quasiparticle state is another instance of the Anderson orthogonality catastrophe phenomenon.

We would like to point out that it is not necessary for the self energy to have the above marginal form for obtaining the $\ln(1/T)$ divergence of the specific heat coefficient. One could always resort to the microscopic models where the abundance of bosonic low-energy excitations below the temperature leads to such singularities when these excitations interact with the electrons. However Eq. (13.1) is very useful for approximately describing the properties

[2] C. M. Varma, P. B. Littlewood, S. Schmitt-Rink, E. Abrahams and A. E. Ruckenstein, *Phys. Rev. Lett.* **63**, 1996 (1989).

of a large variety of systems from a Landau Fermi liquid perspective without going into the microscopic details. For example, marginal Fermi liquid theory has been quite successful in describing the normal state properties of the strange metal regime of high temperature superconductors, correctly accounting for the observed linear-T resistivity, frequency dependence of the optical conductivity, linear-V corrections in the tunneling conductivity, existence of a constant and temperature independent component in the NMR relaxation rate, T_1^{-1}.

We finally discuss possible three-dimensional microscopic models of interacting fermions from which one can obtain a self energy having the form shown in Eq. (13.1) remaining valid down to $T = 0$ over a finite region of parameter space. In Hertz theory, we have seen that the Fermi-liquid behavior prevails at sufficiently low temperatures, $T < |r|^{z/2}$, so strictly speaking, marginal behavior persists to $T = 0$ only at the quantum critical point. Next we consider the case of transverse electromagnetic fluctuations that we studied in Sec. 10.2. At the one loop level, using Eq. (10.18) for the photon propagator, the self energy is given by

$$\Sigma(i\nu_n, \mathbf{k}) = \left(\frac{e}{m}\right)^2 \frac{1}{\beta} \sum_{s;a,b} \int (d\mathbf{q}) \frac{k_a(k_b - q_b)}{i\nu_n - i\omega_s - \xi(\mathbf{k}-\mathbf{q})} \frac{\delta_{ab} - q_a q_b/\mathbf{q}^2}{\frac{1}{\mu}\mathbf{q}^2 + \frac{|\omega_s|}{\eta|\mathbf{q}|}}. \quad (13.4)$$

Assume $\nu_n > 0$ for simplicity. The Matsubara sum over ω_s is carried out first by analytically continuing to complex frequencies ω and replacing the sum by a contour integral that encloses only simple poles of the function $\coth(\beta\omega/2)$, i.e., the bosonic Matsubara frequencies. Note that the integrand has a branch point at $\text{Im}\omega = 0$, so our choice of contour is very much like Fig. 9.7 except that now we deform the contour in the upper half-plane so as to avoid the "unwanted" pole at $i\nu_n - \xi(\mathbf{k}-\mathbf{q})$. The sum over a, b gives $\mathbf{k}^2 - (\mathbf{k}\cdot\mathbf{p})^2 = \mathbf{k}^2 \sin^2\theta_{\mathbf{kq}}$, and we can take $\mathbf{k}^2$ out of the integral. The contour integral vanishes over the big semicircles and we are left with line integrals over the $\text{Re}\omega$ axis and the contribution from going around the aforementioned pole in the upper half-plane. The frequency integration is restricted to a maximum magnitude of $\eta|\mathbf{q}|$ for otherwise the transverse photon modes will not remain overdamped. Finally, the integration over $\mathbf{q}$ is performed. It is not important for us to go into the details of the integration, except to say that it follows the same strategies that we employed for calculating the free energies in Hertz theory and later for the case of transverse electromagnetic fluctuations. At the Fermi surface, and at zero temperature, the self energy behaves as

$$\Sigma(\epsilon, \mathbf{p}_F) \sim -e^2 \mu v_F \epsilon \ln(\epsilon_c/\epsilon), \quad (13.5)$$

in agreement with Eq. (13.1) with $\lambda = e^2\mu v_F$. This result holds as long as the energies satisfy $|\epsilon|/\epsilon_F > (|p - p_F|/p_F)^3$, which is obviously true at the Fermi surface. For smaller energies such that $|\epsilon|/\epsilon_F < (|p - p_F|/p_F)^3$, the logarithmic term goes away and Fermi liquid results are recovered. There is evident similarity with Hertz theory, with the difference that the Fermi surface is analogous to the quantum critical point, $r = 0$, in Hertz theory. Restoring conventional units would give us the size of the dimensionless constant λ determining the strength of the singularity,

$$\lambda \simeq \mu \frac{e^2 \mu_0 c}{h} \frac{v_F}{c}. \tag{13.6}$$

Unless the magnetic permittivity μ happens to be rather large (possible near a ferromagnetic QCP), the effect involves the product of two small quantities, and the marginal behavior may not begin to dominate the Landau Fermi liquid contributions unless one goes to really low energies.

How then do the marginal Fermi liquid effects manifest so beautifully in the high temperature superconductors? One possibility is that in some strongly-correlated systems, the low energy properties may be described by a model that has the form of interacting fermions strongly coupled to some effective transverse "electromagnetic" fields. Furthermore, since these gauge fields are no longer electromagnetic in origin, the effective speed of light may be much smaller, governed by electronic energy scales. It may just be possible to have a microscopic route to a meaningfully significant marginal Fermi liquid behavior.

Index